The Minerals, Metals & Materials Series

Henry A. Colorado Lopera ·
Felipe Perisse Duarte Lopes ·
Sergio Neves Monteiro · Daniel Salazar ·
Marc A. Meyers
Editors

Natural Fibers and Biocomposites 2026

TMS

Springer

Editors
Henry A. Colorado Lopera
Universidad de Antioquia
Medellín, Colombia

Sergio Neves Monteiro
Military Institute of Engineering
Rio de Janeiro, Brazil

Marc A. Meyers
University of California San Diego
San Diego, CA, USA

Felipe Perisse Duarte Lopes
Universidade Estadual do Norte Fluminense
Rio de Janeiro, Brazil

Daniel Salazar
BCMaterials
Leioa, Spain

ISSN 2367-1181 ISSN 2367-1696 (electronic)
The Minerals, Metals & Materials Series
ISBN 978-3-032-13576-6 ISBN 978-3-032-13577-3 (eBook)
https://doi.org/10.1007/978-3-032-13577-3

Cover Illustration: Unpublished photographs provided by Felipe Perisse Durarte Lopes.

This Springer imprint is published by the registered company Springer Nature Switzerland AG
The registered company address is: Gewerbestrasse 11, 6330 Cham, Switzerland

Preface

The reliance of conventional engineered composite materials on synthetic resins and fibers creates a significant environmental concern. Given the continuous rise in demand for high-performance materials, it is critical to find sustainable alternatives that satisfy stringent industry requirements while minimizing their ecological footprint. Natural fibers and biocomposites are a promising solution, as they are derived from renewable biological sources (plants, animals, and microorganisms) and offer advantages like a reduced carbon footprint, lower energy consumption, and improved biodegradability. These natural fibers possess remarkable mechanical properties—such as a high strength-to-weight ratio, stiffness, and toughness—and when combined with suitable bio-based matrices, they form biocomposites that can match or even surpass conventional materials in certain applications. However, the transition to natural fiber biocomposites is challenging, with performance heavily influenced by factors like growth conditions, harvesting, processing techniques, and the interfacing mechanisms needed to achieve optimal mechanical and structural metrics; this challenge is the core motivation and objective of this current effort.

This proceedings volume offers a comprehensive overview of recent advancements in the processing and characterizing of natural fibers and biocomposites, highlighting their potential as sustainable solutions to mitigate environmental challenges, and showing their applications. Natural fibers, derived from renewable sources, possess exceptional properties, biodegradability, and low cost, making them a compelling alternative to synthetic fibers in composite reinforcement and other applications. The chapters in this book underscore the growing research and industrial interest in natural fibers to reduce the carbon footprint associated with traditional synthetic materials, given their ability to absorb carbon dioxide during growth. The imperative to develop sustainable, biodegradable, or biocompatible solutions has spurred applications in diverse fields, including sports, transportation, armor, medicine, infrastructure, construction and building materials, and architecture.

Henry A. Colorado Lopera

Contents

About the Editors

Henry A. Colorado Lopera is a Full Professor in the Mechanical and Materials Engineering Departments of the Universidad de Antioquia at Medellin, Colombia. He is also affiliated with the Materials Department from the State University of Northern Rio de Janeiro, Brazil and has been a Visitor Professor in the Military Institute of Engineering and a Fulbright Visitor Scholar in the Mechanical and Aerospace Engineering Department of the University of California San Diego from 2023 to 2024. His main areas of research are composite materials, natural fibers, green materials, and additive manufacturing. Dr. Lopera obtained a BS and M.Sc. in Mechanical Engineering from Universidad Nacional de Colombia in 2005; and M.Sc. and Ph.D. in Materials Science and Engineering from the University of California Los Angeles in 2010 and 2013, respectively. Henry has active collaborations in several institutions in Brazil, Chile, China, and the United States. He is also working as Associate Editor for the journals *Heliyon*, *Engineered Science*, and *ES Materials & Manufacturing*; and Editorial Board for *Advanced Composites and Hybrid Materials* and for the *American Ceramics Society Bulletin* (2023–2026). He is also a Global Ambassador for the American Ceramics Society.

Felipe Perisse Duarte Lopes has a degree in Materials and Metallurgical Engineering from the Darcy Ribeiro State University of Northern Fluminense (2009), a master's degree in Materials Engineering and Science from the Darcy Ribeiro State University of Northern Fluminense (2011), and a Ph.D. in Materials Science from the Military Institute of Engineering (2013) completed in a sandwich program in partnership with the University of California, San Diego. He also completed a post-doctorate at the Advanced Materials Laboratory at UENF (2018–2023). Dr. Lopes is currently an Associate Professor at LAMAV/UENF. He has experience in materials and metallurgical engineering, with an emphasis on mechanical properties of polymeric composites, metals, and alloys, working mainly on the following topics: polymeric composites, natural fibers, mechanical properties, thermal properties, failure analysis, corrosion and inspection of equipment in offshore units. Dr. Lopes worked for over 10 years as an inspection and maintenance engineer in the oil and gas industry around the world, in countries such as the United States, Norway, England, Spain, Singapore, and Mexico.

Sergio Neves Monteiro graduated as a Metallurgical engineer (1966) at the Federal University of Rio de Janeiro (UFRJ) Brazil. He received his M.Sc. (1967) and Ph.D. (1972) from the University of Florida, USA, followed by a 1975 course in energy at the Brazilian War College, and a post doctorate (1976) at the University of Stuttgart, Germany. In 1968, he joined the Metallurgy Department of UFRJ as full professor of the post-graduation program in engineering (COPPE). He was elected as head of department (1978), coordinator of COPPE (1982), Under-Rector for Research (1983), and was invited as Under-Secretary of Science for the State of Rio de Janeiro (1985), and Under-Secretary of the College Education for the Federal Government (1989). He retired in 1993 from the UFRJ and joined the State University of North Rio de Janeiro (UENF), where he retired in 2012. He is now a voluntary professor at the Military Institute of Engineering (IME), Rio de Janeiro. Dr. Monteiro has published about 2,500 articles in journals and conference proceedings with more than 15,600 citations in Scopus (H index 63) and 24,700 in Google Scholar (H index 78). He has been honored with several awards including the ASM Fellowship, the Brazilian Army Medal, the Humbertus Colpaert Medal, and several TMS awards, like the Leadership Award (2020), as well as the 2024 Fray International Award. He is a top researcher (1A) of the Brazilian Council for Scientific and Technological Development (CNPq) and Emeritus Scientist of State of Rio de Janeiro (FAPERJ). He was president of the Superior Council of the State of Rio de Janeiro Research Foundation, FAPERJ (2012), and coordinator of the Engineering Area of this foundation (2016). Dr. Monteiro has also served as president of the Brazilian Association for Metallurgy, Materials and Mining (ABM, 2017–2019), as a consultant for the main Brazilian R&D agencies, and as a member of the editorial board of five international journals as well as associate editor-in-chief of the *Journal of Materials Research and Technology*. He is the author of 130 patents and a top world researcher in "Natural Fiber Composites" and "Ballistic Armor", Scopus 2020. He has contributed to the Graduate Program of Materials Science and Engineering to reach the top position in the Brazilian Federal Agency for Support and Evaluation of Graduate Education (CAPES) institutional rank.

Daniel Salazar is an electronic engineer from Universidad del Quindío (2007) and earned his doctoral degree in Physics and Mathematics from Universidad de Castilla-La Mancha in 2012. During his Ph.D., he researched nanostructured and granular magnetic materials, magnetic properties and electrical transport in nanocrystalline oxides, and the magnetic, structural, and thermal properties of strongly correlated electron systems. Since late 2013, he has been at BCMaterials—initially as a postdoctoral researcher and currently as a Senior Researcher. His work focuses on the magnetocaloric effect and its underlying physics in single crystals of magnetic shape-memory alloys, as well as enhancing coercivity in nanostructured rare-earth-free hard magnets through grain-boundary engineering. Recently, he has begun developing formulations for multifunctional magnetic inks, pastes, and powders as feedstock materials for 3D-printing processes. Dr. Salazar has co-supervised three doctoral theses (nine more in progress), seven master's theses, and several bachelor's theses. He has published more than 65 SCI papers and four book chapters (h-index 19; i10-index 34; 1,122 citations). He has contributed to over 50 international conferences, delivered twenty invited talks at international conferences, and given numerous seminars at institutions including Oak Ridge National Laboratory (USA), Lawrence Livermore National Laboratory (USA), Ames Laboratory (USA), TU Darmstadt (Germany), University of Pittsburgh (USA), NIMS and TIT (Japan), and Texas A&M University (USA). He has been a visiting scientist at NCSR-Demokritos (Greece), the University of Delaware (USA), and Universidad del Valle (Colombia). He has participated in more than 15 publicly funded projects and one privately funded project, including those supported by the EU Horizon (H2020) program and the European Space Agency. He has organized over 15 international symposia/workshops and serves on the boards of five Scientific Advisory Committees. He is the former Chair of the Magnetic Materials Committee and a former member of the Scientific Program Board of The Minerals, Metals & Materials Society (TMS, USA). He is also a member of the Spanish Club of Magnetism, the European Magnetism Association, SOCIEMAT, and E-MRS.

Marc A. Meyers is a graduate (Mechanical Engineering) of the Federal University of Minas Gerais and did his doctorate at the University of Denver. He held university positions at the Military Institute of Engineering (Rio de Janeiro), South Dakota School of Mines and Technology, and New Mexico Institute of Mining and Technology, prior to joining the University of California, where he is Distinguished Professor in San Diego.

In the educational area, taught over hundred classes, advised 57 doctoral dissertations and authored or co-authored four successful textbooks translated into Chinese and Portuguese: Meyers and Chawla, *Mechanical Metallurgy* (Prentice Hall, 1983); Meyers and Chawla, *Mechanical Behavior of Materials* (Cambridge, 2004; 2024); Meyers, *Dynamic Behavior of Materials* (J. Wiley, 1994); Meyers and Chen, and *Biological Materials Science* (Cambridge, 2014). As Associate Director and Director of the NSF-funded Institute for Mechanics and Materials, he organized four summer schools with leading scholars and students recruited globally and co-edited the book: Meyers, Armstrong, and Kirchner, eds. *Mechanics and Materials* (J. Wiley, 1999).

He has had extended stays at laboratories in the UK (Cambridge), Japan (Tsukuba), Germany (Karlsruhe and Saarbrucken), France (U. Lorraine), and China (Institute for Metal Research). He been recognized by accolades in the United States, Europe, China, and Brazil. In the United States, he received awards from the three societies in which he is most active: TMS, APS, and ASM International; he is a fellow of all three. Within ASM, he received the Albert Easton White Teaching Award (nominated by the local chapter) and the Colorado School of Mines Barrett Silver Medal.

Part I
Sustainable and Functional Composites and Biocomposites

Characterization of Polymer Composites Reinforced with Coconut Sheath Fibers

João Victor Chaim Almeida, David Coverdale Rangel Velasco, Victor Paes Dias Gonçalves, Michel Picanço Oliveira, Noan Tonini Simonassi, Carlos Maurício Fontes Vieira, and Felipe Perissé Duarte Lopes

Abstract The growing demand for sustainable alternatives to synthetic fibers has driven interest in natural fibers as reinforcing agents in polymer composites. In this study, coconut sheath fiber (CBF) was employed as reinforcement in a polymer matrix, with volumetric fractions ranging from 10 to 30%. The fibers were extracted from coconut palm sheaths, and their orientation and distribution were considered during specimen preparation. The impact performance of the composites was evaluated through Izod impact testing (ASTM D256). The results revealed a nonlinear trend, with energy absorption increasing significantly up to 25% fiber content, followed by a decrease at 30%, likely due to processing-induced defects and poor fiber dispersion. These findings highlight the potential of CBF as a sustainable reinforcement, while also indicating the existence of an optimal fiber content for maximizing impact resistance.

Keywords Natural fabric · Coconut sheath fiber · Natural fibers · Polymer composites · Mechanical test · Izod impact test

Introduction

Natural fibers come in a wide variety, and due to the expansion of their applications, there is a growing need to utilize them in the field of engineering [1]. Researchers have been seeking sustainable alternatives to replace synthetic fibers, and natural fibers (NF) are being studied as reinforcements in polymer composites. These fibers can be obtained from different parts of plants, such as seeds, fruits, leaves, and stems. Among the most common examples are cotton, coconut, flax, and bamboo [2]. In addition, because they are biodegradable, they are classified as renewable fibers [3].

J. V. C. Almeida (✉) · D. C. R. Velasco · V. P. D. Gonçalves · M. P. Oliveira · N. T. Simonassi · C. M. F. Vieira · F. P. D. Lopes
State University of the Northern Rio de Janeiro—UENF, Advanced Materials Laboratory—LAMAV, Campos Dos Goytacazes, Brazil
e-mail: joaovictorchaim@gmail.com

H. A. C. Lopera et al. (eds.), *Natural Fibers and Biocomposites 2026*, The Minerals, Metals & Materials Series, https://doi.org/10.1007/978-3-032-13577-3_1

The intense environmental degradation that occurred in the last century, largely due to rapid industrialization, has driven the search for alternatives capable of mitigating its effects. With the strengthening of environmental legislation, researchers have directed their studies toward the development of sustainable materials. In this context, plant fibers emerge as a viable option to replace synthetic materials [4]. In recent years, the use of natural fibers has shown significant advances, becoming the subject of increasingly frequent research, with a view to their widespread application [5].

Because they are naturally available, some fibers have shapes and dimensions that cannot be modified [2]. Factors such as fiber length and orientation play a key role in assessing their mechanical properties. In addition, each type of fiber has its own structural characteristics that need to be taken into account [4]. Another advantage is the favorable cost–benefit ratio of using natural fibers, since they are low-density materials that are widely available in nature and inexpensive.

The NF used in this study acts as a polymer reinforcement, with epoxy resin as the matrix material. Cocos nucifera, commonly known as coconut palm, is widely found in Brazil. The fiber used corresponds to the natural tissue present in the coconut husk, located between the petiole and the trunk of the plant [4], as shown in Fig. 1, the tree and the fiber after being removed. Structurally, this fiber is formed by a core with a larger diameter, surrounded and intertwined by external fibers [6].

Previous research on the use of different parts of the coconut, such as leaves and sheath, is widely found in the literature. However, the study of coconut sheath tissue

(a) (b)

Fig. 1 Location of the sheath on the coconut palm and the coconut sheath tissue represents a more recent line of research, with few studies available. The purpose is to evaluate its properties through mechanical testing. In general, existing studies focus on fibers with a random distribution matrix, while the coconut sheath has a natural formation that directly influences the results obtained. The fiber has an angle of approximately 45°, where its fabric morphology has constant fibers. From this, it is possible to identify sides with a smaller angle (greater resistance) and a larger angle (lesser resistance). It is believed that the smaller the proposed angle, the greater the resistance of the fabric in that direction. Therefore, based on the manufacture of the "Raw" Test Specimen in a matrix, it is possible to decide whether to use it to test the side with greater or lesser resistance. Some important details are that the orientation of the fiber must be adjusted considering the post-cutting/manufacturing of the specimens to be tested. The arrangement of the fibers is decisive for performance, with regions of greater and lesser mechanical resistance. For the preparation of the samples, we sought to prioritize the greatest possible resistance, using the entire fabric in order to achieve the best result [7].

Materials and Methods

Materials

The employed material was the vegetal tissue found between the petiole and the trunk of the coconut palm (*Cocos nucifera*). The samples were obtained via manual extraction and delivered unprocessed (*in natura*). For the fabrication of the specimens, the epoxy resin Diglycidyl Ether of Bisphenol A (DGEBA) was utilized. This resin was combined with the curing agent (hardener) diethylenetriamine (DETA), both supplied by Avipol and manufactured by Silaex. The resulting mixture established the polymeric matrix.

Methods

The reinforcement material was initially cut at specific angles corresponding to the dimensions of the final mold (matrix) in which it would be used, as shown in Fig. 2a. After cutting, the material was washed with running water to remove any impurities from the surface. The clean fabric was then transferred to an oven for drying at 60 °C for 24 h. In addition, fiber samples were cut at varying angles to facilitate subsequent mechanical testing focused on the dependence of properties on angulation. The samples were manufactured using a closed mold process. Before use, the cleanliness of each mold component was verified, and a release agent was applied to all internal surfaces. A layer of plastic film was then placed on the lower and upper sections of the mold to aid in the demolding process. The epoxy resin system

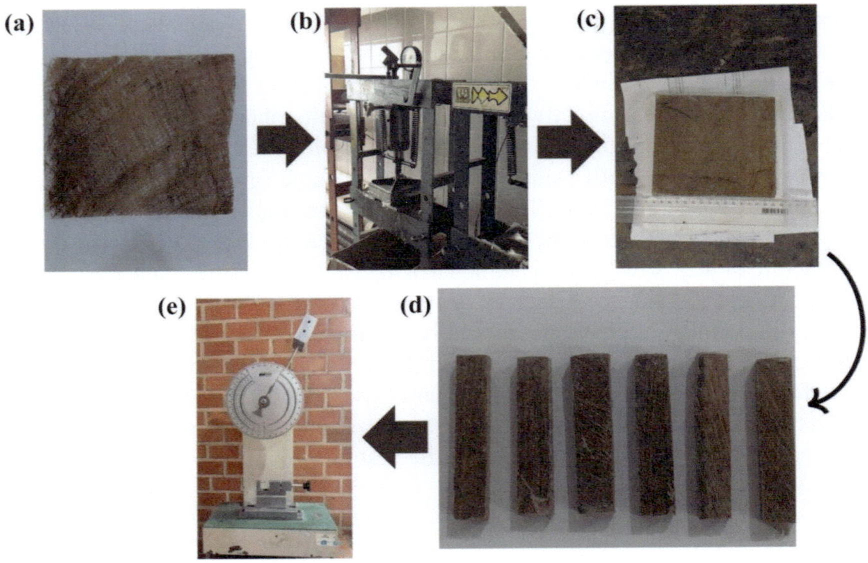

Fig. 2 **a** Coconut sheath fiber; **b** matrix under pressure during the making of the plates, **c** composite plate after pressing, **d** test specimen after cutting and pre-notching and **e** PANTEC brand XC-50 impact machine

(resin and hardener) was mechanically mixed for 3 to 5 min, or until an observable exothermic increase in temperature occurred. The composite plate was manufactured by manual lamination, reaching target volumetric fractions (Vf) of 10%, 15%, 20%, 25%, and 30%. Figure 2b shows the assembled mold subjected to a hydraulic press, where a pressure of 5 tons was applied for a total period of 24 h.

After demolding, the composite plates, illustrated in Fig. 2c, were cut with a band saw to obtain the initial samples. These samples were then measured (thickness, width, and length) and sanded to remove excess resin and achieve dimensional conformity, as shown in Fig. 2d. The final samples for testing were prepared in accordance with ASTM D256-24. The hardness of the samples was determined using a Shore D durometer. Impact resistance was evaluated using the Izod test (ASTM D256). Notched rectangular samples measuring 80 × 10 × 4 mm were positioned vertically and impacted in a pendulum impact tester, shown in Fig. 2e. A minimum of five samples per volumetric fraction were tested, and the absorbed energy (J/m) was recorded, with mean values and standard deviations calculated for comparative analysis.

Results and Discussion

The Izod impact test results shown in Table 1 reveal a nonlinear relationship between the volumetric fraction of reinforcement and the absorbed impact energy. At lower reinforcement contents (10–15%), only a slight increase in impact resistance was observed, suggesting that the fibrous or particulate phase was not yet sufficient to promote a significant hardening effect. A marked improvement occurred at 20% and 25%, where the absorbed energy increased sharply, reaching a maximum value of 0.23 J at 25%. This behavior indicates that, within this range, the reinforcement effectively contributed to energy dissipation mechanisms, such as crack deflection, fiber pull-out, and interfacial delamination, which are commonly associated with improved impact performance in composite systems.

Figure 3 illustrates the variation of Izod impact resistance as a function of volumetric fraction. A progressive increase in impact resistance was observed from 10 to 25%, with the maximum value recorded at 25% (≈0.23 J/m). This trend indicates that, up to this concentration, the reinforcement phase plays a significant role in dissipating impact energy through mechanisms such as crack deflection, fiber pull-out, and interfacial debonding. At 30% volumetric fraction, however, the impact resistance decreased markedly, suggesting the existence of a critical threshold beyond which the addition of reinforcement is detrimental. This reduction can be attributed to defects introduced during processing, such as voids or poor dispersion, as well as possible agglomeration of the reinforcing phase. These factors act as stress concentrators, compromising the efficiency of stress transfer and thus limiting the material's ability to absorb impact energy. The presence of error bars in Fig. 3 also highlights variability in the mechanical response, particularly at intermediate fractions (20–25%), where the largest deviations were observed. This variability may result from microstructural heterogeneities inherent to composite manufacturing. Nevertheless, the overall trend supports the conclusion that 25% represents the optimal volumetric fraction for maximizing impact resistance under the conditions tested.

The Shore D hardness test results shown in Table 2 and its graph in Fig. 4 demonstrate a consistent trend across all volumetric fractions evaluated. The highest hardness value was observed at 10% (82.5), suggesting that, with a lower reinforcement content, the matrix maintained greater structural continuity, thus contributing to greater resistance to localized deformation.

Table 1 Results of Izod test X fraction volumetric

Impact Izod test (J)	Fraction volumetric (%)
0,044 J	10
0,052 J	15
0,17 J	20
0,23 J	25
0,12 J	30

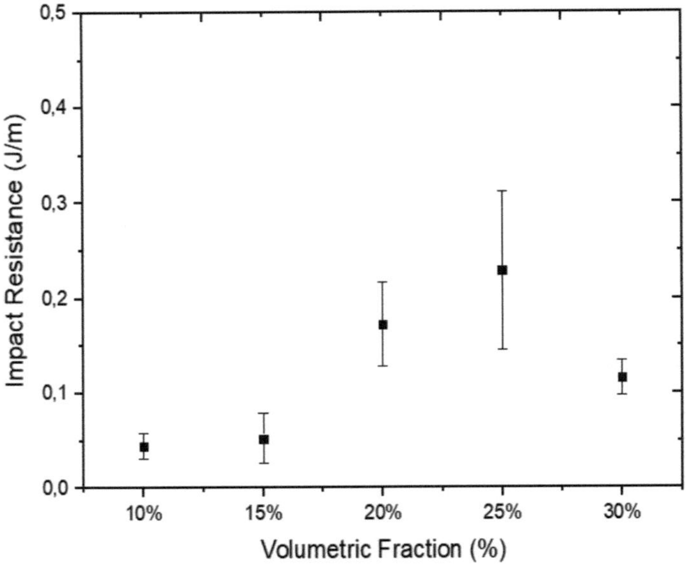

Fig. 3 Impact testing results by Izod configuration of composites studied

Table 2 Results of hardness test

Hardness test shore D	Volumetric fraction (%)
82,50 (±0,10)	10
73,75 (±0,35)	15
75,75 (±0,35)	20
73,75 (±1,77)	25
69,00 (±3,54)	30
74,00 (±0,71)	35

As the reinforcement fraction increased to the range of 15–25%, the hardness values stabilized around 73–76, with relatively low standard deviations, indicating a uniform mechanical response. This behavior suggests that the reinforcing phase reached a threshold where its contribution to hardness was balanced by a reduction in the continuous matrix fraction, leading to a plateau effect in the measured values. At 30%, a noticeable decrease in hardness was observed (69 ± 3.54), accompanied by greater variability. This reduction may be associated with microstructural defects, such as voids or agglomerates, which compromise the compactness of the composite and hinder efficient stress distribution under indentation. Interestingly, at 35% the hardness slightly increased again (74 ± 0.71), which could be related to local densification or improved packing of the reinforcement at higher fractions, partially compensating for the matrix discontinuities.

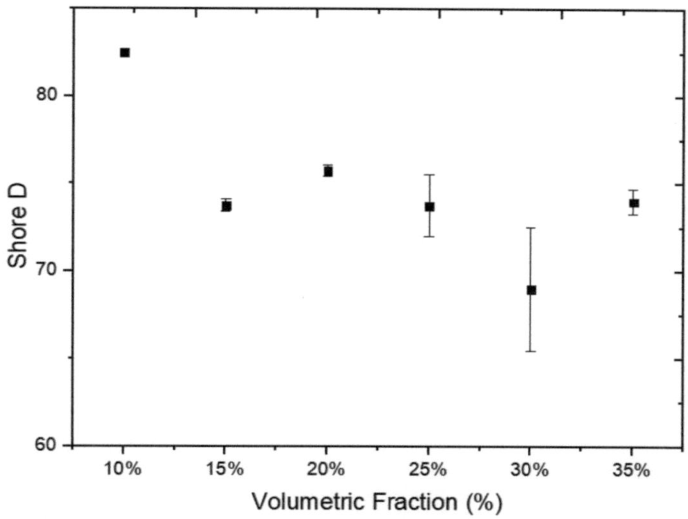

Fig. 4 Hardness testing results by Shore D configuration of composites studied

Overall, the results suggest that the reinforcement content exerts a significant influence on hardness, with the 10% fraction delivering the best performance. However, within the range of 15–35%, the values tend to converge, indicating that beyond a certain threshold, the contribution of additional reinforcement does not proportionally enhance hardness and may even introduce heterogeneity.

This behavior is attributed to the anisotropic nature of laminated composites, where the load applied by the indenter interacts differently with the reinforcement–matrix interfaces. When the indentation is aligned with the fiber plane, the reinforcement provides greater resistance to localized deformation, increasing the surface hardness. On the other hand, when tested perpendicular to the lamination plane, the indenter mainly involves resin-rich regions, resulting in slightly lower hardness values. These findings emphasize that the hardness response of the composite is direction-dependent, reflecting the intrinsic heterogeneity introduced by the manual lamination process.

Conclusions

Therefore, it should be noted that not only the volume fraction, but also the location where the notch will be made, are important factors to consider during manufacturing. The best results were obtained with a volumetric fraction of 25%; however, the 20% fraction has a lower standard deviation, which can also influence the decision on values. It can also be noted that the 10% fraction has a higher hardness value than the other fractions, which impacts material savings for a composite with greater hardness.

Acknowledgements The authors thank the Brazilian agencies CNPq, process number 302976/2022-1, and FAPERJ, process number E-26/200.847/2021.

References

1. Nadendla S (2018) Fabrication of chemically treated pure-bred coconut leaf sheath fibre reinforced composites and determination of mechanical properties. Macromol Symp 382:1800119. https://doi.org/10.1002/masy.201800119
2. Aruchamy K, Karuppusamy M, Krishnakumar S, Palanisamy S, Jayamani M, Sureshkumar K et al (2025) Enhancement of mechanical properties of hybrid polymer composites using palmyra palm and coconut sheath fibers: The role of tamarind shell powder. BioResources 20(1):698–724
3. da Silveira PHPM, Santos MCC, Chaves YS, Ribeiro MP, Marchi BZ, Monteiro SN et al (2023) Characterization of thermo-mechanical and chemical properties of polypropylene/hemp fiber biocomposites: Impact of maleic anhydride compatibilizer and fiber content. Polymers 15:3271. https://doi.org/10.3390/polym15153271
4. Bharath KN, Mavinkere Rangappa S, Siengchin S, Alexey K, Gorbatyuk S (2022) Mechanical characterization of natural fiber reinforced polymer composites. Polym Compos. 43(4):1985. https://doi.org/10.1002/pc.26513
5. Valášek P, Müller M, Šleger V, Kolář V, Hromasová M, D'Amato R et al (2021) Influence of alkali treatment on the microstructure and mechanical properties of coir and abaca fibers. Materials 14:2636. https://doi.org/10.3390/ma14102636
6. Khan T, Karthikeyan N, Naveen J, Anand SP, Sebaey TA (2024) Coconut (*Cocos nucifera*) sheath-based polymeric composites - A review. Heliyon. 10(15):e35644
7. Almeida JVC, Rangel Velasco DC, Simonassi NT, Fontes Vieira CM, Duarte Lopes FP (2025) Characterization of composite polymers reinforced with coconut sheath fibers. In: Lopera HAC, et al., editors. Natural fibers and biocomposites. Springer, Cham. https://doi.org/10.1007/978-3-031-81190-6_11

Effect of Açaí Palm Stem Fiber Content on the Properties of Polyester-Based Composites

Leonardo Martins Nazário Silva, Marcela Coelho da Silva, Heliane Rosa do Amaral, Felipe Perissé Duarte Lopes, Hemanuelly Ferreira Breda Lan Oliveira, Felipe Gabriel Santos Araújo, Henry Alonso Colorado Lopera, and Michel Picanço Oliveira

Abstract Lignocellulosic fibers are increasingly employed in various industrial sectors due to their low cost, mechanical strength, and biodegradability. This study investigates polyester matrix composites reinforced with different concentrations (10%, 20%, and 30%) of fibers from the stem of Euterpe oleracea (açaí palm). The composites were characterized through Charpy impact and tensile tests, Fourier-transform infrared spectroscopy (FTIR), thermogravimetric analysis (TGA), and scanning electron microscopy (SEM). The incorporation of açaí fibers substantially enhanced the mechanical performance, with increases of up to 2301% in impact resistance, 423% in ultimate tensile strength, 111% in elongation at break, 432% in elastic modulus, and 1004% in toughness. Conversely, thermal stability decreased, with the peak degradation temperature reduced from 382 °C for neat polyester to 332–373 °C for the fiber-reinforced composites. SEM images revealed weak fiber–matrix adhesion, likely due to the presence of surface waxes and natural impurities, although

L. M. N. Silva · M. C. da Silva · H. R. do Amaral · H. F. B. L. Oliveira · F. G. S. Araújo · M. P. Oliveira (✉)
Forest and Wood Sciences Department, Federal University of Espírito Santo, Jeronimo Monteiro, Brazil
e-mail: michel.oliveira@ufes.br

L. M. N. Silva
e-mail: leonardo.mn.silva@edu.ufes.br

M. C. da Silva
e-mail: marcela.c.silva@edu.ufes.br

F. G. S. Araújo
e-mail: felipe.g.araujo@edu.ufes.br

F. P. D. Lopes
Materials Science and Engineering Department, State University of Northern Rio de Janeiro, Campos Dos Goytacazes, Brazil

H. A. C. Lopera
CCComposites Lab., Universidad de Antioquia, Antioquia, Colombia
e-mail: henry.colorado@udea.edu.co

H. A. C. Lopera et al. (eds.), *Natural Fibers and Biocomposites 2026*, The Minerals, Metals & Materials Series, https://doi.org/10.1007/978-3-032-13577-3_2

the fibers effectively impeded crack propagation, enhancing toughness. These findings demonstrate that açaí palm fibers are promising reinforcements for polyester composites, improving mechanical behavior while slightly compromising thermal stability. The results highlight the potential of lignocellulosic residues as sustainable alternatives for developing high-performance, eco-friendly composite materials.

Keywords Lignocellulosic fibers · Polyester composites · *Euterpe oleracea*

Introduction

The continuous evolution of the industrial sector has intensified the demand for advanced materials that simultaneously satisfy stringent performance criteria and contribute to atmospheric-pollution mitigation, in accordance with the Sustainable Development Goals (SDGs). In response, research has focused on materials that offer a balanced combination of mechanical strength, low density, cost-effectiveness, and intrinsic sustainability. Within this framework, fiber-reinforced polymer composites have emerged as a particularly promising class of materials [1].

Polymer-matrix composites reinforced with continuous fibers are traditionally fabricated using synthetic reinforcements (e.g., glass, Kevlar®, aramid, polyester), which provide low density, high tensile strength, and elevated elastic modulus [2]. Natural fibers are emerging as a sustainable reinforcement option for renewable biocomposites. These fibers, favored over synthetic alternatives, offer biodegradability, lightweight properties, and renewability [3]. They exhibit favorable mechanical characteristics, low damage susceptibility, and excellent thermal and acoustic insulation [4]. Furthermore, natural fibers are non-irritating, readily available, cost-effective, and have a low embodied energy footprint. Their use also presents advantages in processing, with non-abrasive qualities and neutral CO_2 emissions, while potentially boosting economic opportunities in rural areas [3–8].

The açaí palm (*Euterpe oleracea*) is native to the Amazon region in Brazil's Northern region which is a significant crop supporting approximately 25,000 direct and indirect jobs. As of 2021, açaí cultivation covered 208,000 hectares, a substantial increase of 72,000 hectares since 2015 [9, 10]. Developed palms (15–20 years old) can exceed 15 m in height and can present harvesting challenges that sometimes lead to stem extraction. These stems yield fibers that are utilized in composite reinforcement applications [11].

In this context, the present study developed and rigorously characterized polyester-matrix composites reinforced with açaí-palm-stem fibers at 0, 10, 20, and 30 wt %. Standardized specimens were fabricated and subjected to tensile and Charpy impact tests to evaluate mechanical performance. Complementary thermal analyses clarified how fiber loading affected composite behavior. Scanning electron microscopy (SEM) was employed to examine interfacial morphology and elucidate fracture mechanisms. By combining these methods, we aimed to validate a high performance of the composite produced and to advance the scientific understanding of natural-fiber-reinforced polymer systems.

Experimental

Materials

Açaí palm (Euterpe oleracea) stems were harvested by a local producer in the southern region of Espírito Santo State, near the city of Rio Novo do Sul, Brazil. The fibers were manually extracted from the outer layer of the stems, exhibiting diameters ranging from 0.5 to 1.8 mm and an average aspect ratio of approximately 60–100, depending on the extraction region. After extraction, the fibers were washed with distilled water to remove impurities, air-dried at room temperature for 48 h, and subsequently oven-dried in a SolidSteel sterilization and drying oven at 60 °C for 24 h to prevent fungal and mold growth. Unsaturated polyester resin ARAZYN 25,108 T-10 and methyl ethyl ketone peroxide catalyst (Butanox M-50) were used as received.

Composite Production

A carbon-steel mold was fabricated according to ASTM D6110-18 (Charpy impact test) and ASTM D3039-17 (tensile test) specifications. The polyester resin was catalyzed with 1.5 wt% Butanox M-50 and mixed until homogenized. Predried lignocellulosic fibers were weighed to achieve filler contents of 0%, 10%, 20%, and 30 wt%, corresponding to the samples P0, P10, P20, and P30, respectively. The fibers were placed unidirectionally inside the mold and impregnated manually with resin under slight vacuum to minimize voids.

The laminates were cured under 4.61 MPa pressure for 24 h at 25 ± 2 °C, followed by post-curing at ambient conditions for 48 h. After polymerization, the panels were machined according to the dimensional tolerances prescribed by the ASTM standards and visually inspected to ensure uniform fiber dispersion (Fig. 1).

Characterizations

For each composition (P0, P10, P20, and P30), seven (7) specimens were tested for each mechanical property (impact and tensile), in accordance with ASTM D6110 (2018) and ASTM D3039 (2017), respectively.

(a) (b) (c) (d)

Fig. 1 Composites produced from **a** neat polyester, **b** polyester with 10% filler content, **c** polyester with 20% filler content, and **d** polyester with 30% filler content

Impact Test: Grooves were generated using a CEAST Notchavas Sculptor (2.5 mm depth), and the tests were carried out using a Pantech XJ Series impact tester equipped with an 11 J pendulum hammer operating at 2.9 m/s.

Tensile Test: Performed using an EMIC DL 10,000 testing machine with a 5 kN load cell and a crosshead speed of 2 mm/min until specimen failure.

Fourier-Transform Infrared (FTIR) Analysis

Fourier-transform infrared (FTIR) spectra were recorded in attenuated total reflectance (ATR) mode using a Bruker Tensor 27 spectrometer. The spectra were collected over the 4000–600 cm^{-1} range with a resolution of 4 cm^{-1}, providing chemical insights into fiber–matrix interactions and possible modifications resulting from fiber incorporation.

Scanning Electron Microscopy (SEM)

Morphological analysis of the fracture surfaces was carried out using a JEOL JSM-IT200 scanning electron microscope at an accelerating voltage of 10 kV. Samples were fixed on aluminum stubs and sputter-coated with a thin layer of gold prior to imaging.

Thermogravimetric Analysis (TGA)

Thermogravimetric (TGA) and derivative thermogravimetric (DTG) analyses were conducted on a Setaram LabSys Evo system under nitrogen atmosphere from 25 °C to 700 °C at a heating rate of 10 °C/min.

Results and Discussion

The results of the Charpy impact resistance, ultimate tensile strength, strain at break, Young's modulus, and toughness tests performed on polyester (P0) and polymer composites P10, P20, and P30 are presented in Fig. 2.

Based on the graph (Fig. 2a), it can be observed that the incorporation of 10%, 20%, and 30% fiber into the composite matrix resulted in a substantial enhancement of impact resistance, as evidenced by the increase in energy absorption during testing. Compared to the unreinforced composite (0%), the sample with 30% fiber exhibited the most significant improvement, achieving an increase of approximately 2301% in energy absorption. Similarly, the composites with 20 and 10% fiber demonstrated enhancements of 1769 and 796%, respectively, when evaluated against the base polymer. These results align with the findings of Rahman et al. [12], who reported significant gains in impact absorption through the incorporation of lignocellulosic fibers into the polymer matrix. Figure 2b demonstrates the progressive enhancement of maximum stress as a function of fiber content in the polyester-based composites. The maximum stress values significantly increased by 109, 203, and 423% for

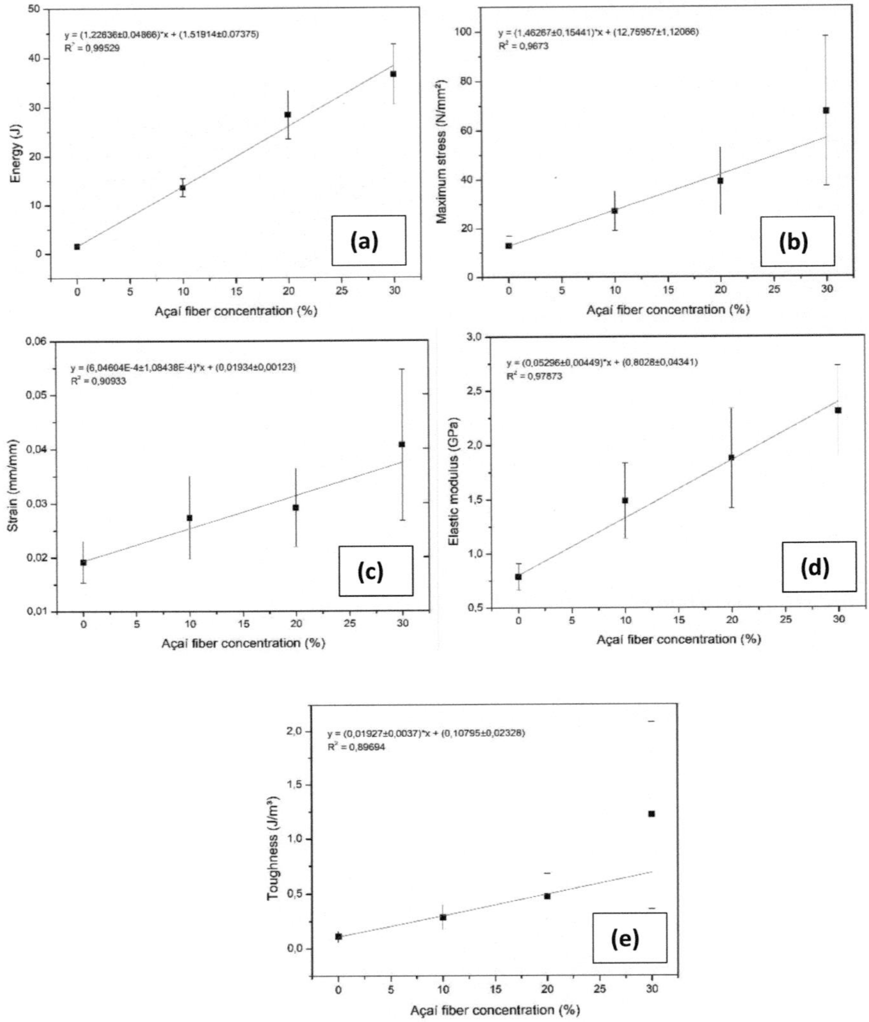

Fig. 2 Comparative analysis of **a** Charpy impact resistance, **b** ultimate tensile strength, **c** strain at break, **d** Young's modulus, and **e** toughness for neat polyester (P0) and polymer composites with 10% (P10), 20% (P20), and 30% (P30) filler content

composites P10, P20, and P30, respectively. This systematic stress increment clearly illustrates the direct correlation between fiber content and mechanical performance [1].

The present study reveals that fiber-reinforced composites exhibited superior plasticity compared to the unreinforced matrix (Fig. 2c). Notably, the composites containing 10, 20, and 30% fiber content demonstrated significant increases in ductility of 42, 52, and 111%, respectively, suggesting a substantial modification of the material's mechanical behavior through fiber reinforcement. Figure 2d reveals a remarkable enhancement in elastic modulus, indicating excellent interfacial interaction between the polyester matrix and fiber reinforcement. The elastic modulus increased dramatically, reaching up to 432% for the P30. These findings corroborate the research by Anuar et al. [13]. Who observed a similar trend, demonstrating that composites with increased fiber content consistently exhibit superior mechanical properties, with higher fiber volumes directly correlating to improved stress transfer and elastic modulus. Toughness exhibited a similar trend. Chirayil et al. [14]. Specifically, the composites P10, P20, and P30 showed increases in toughness of 159, 327, and 1004%, respectively.

Finally, the observed variability in standard deviation can be attributed to inherent process limitations in composite compression molding. During the molding process, the fluidity of the viscous resin under pressure induces fiber migration within the mold cavity. This phenomenon results in a non-uniform fiber distribution, leading to localized heterogeneity in fiber concentration within the composite material. Consequently, regions may exhibit either high or low fiber density, which directly impacts the mechanical performance and contributes to the significant standard deviation observed in experimental results.

Figure 3 presents the TGA and DTG curves, revealing the thermal degradation behavior of the polyester resin and the resulting composites.

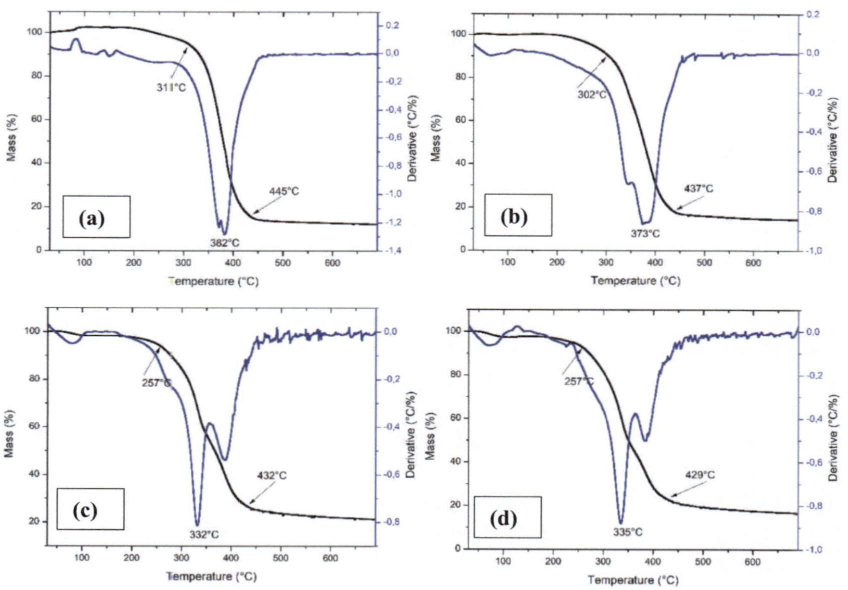

Fig. 3 Comparative thermal analysis for the neat polyester **a** and its composites **b** P10, **c** P20, and **d** P30

Figure 3a–d presents the thermogravimetric analysis (TGA and DTG) of polyester resin and composites with varying filler content (10, 20, and 30%). A slight initial mass loss, likely due to moisture, was observed between 40and 130 °C, particularly in the P20 and P30 composites. Up to 200 °C, all samples showed good thermal stability. Beyond this temperature, degradation started, accelerating between 300 and 429 °C, which corresponds to the decomposition temperatures of hemicellulose, cellulose, and lignin as reported in the literature [15]. DTG analysis indicated that the addition of filler reduced the thermal resistance of the polyester. The neat resin exhibited a peak degradation temperature of 382 °C, while the P10, P20, and P30 composites showed lower peak temperatures of 373 °C, 332 °C, and 335 °C, respectively.

Figure 4a and b illustrates that the neat polyester resin composites exhibited multiple cracks, leading to low resistance due to crack propagation and the absence of rigid particles to enhance stiffness [16]. On the other hand, Fig. 4c demonstrates that the incorporation of 30% lignocellulosic fiber from the açaí stem into the polyester resin composites promoted greater energy absorption. The fibers acted as barriers, inhibiting or deflecting crack propagation [16, 17] and improving toughness through plastic deformation. This mechanism attenuates stress concentration at the crack tip, enabling the composites to withstand higher loads prior to failure [18]. These observations are further supported by the mechanical testing results presented in Fig. 2a–e.

As interfacial interaction is a known determinant of composite mechanical properties [19], the features in Fig. 4d point to poor fiber-matrix adhesion. This weak interfacial adhesion could be related to the presence of naturally occurring substances, such as waxes and mucilage, on the surface of the lignocellulosic fibers [20].

Conclusion

This study demonstrates that incorporating açaí stem fibers into composites significantly enhances mechanical performance. Specifically, the Charpy impact tests indicated an energy absorption increase of up to 2301%, while tensile tests revealed improvements of up to 423% in ultimate tensile strength, 111% in yield strength, 432% in Young's modulus, and 1004% in toughness. These results underscore the fibers' capacity not only to reinforce the composite structure but also to enhance energy dissipation under load, which is crucial for applications demanding high impact resistance. However, thermal analysis indicates a reduction in thermal stability, with the maximum degradation temperature dropping from 382 °C for the neat polymer to values between 332 °C and 373 °C for composite produced. Additionally, SEM image reveals a weak interface between the fibers and the polyester matrix, suggesting that further improvements in fiber-matrix compatibility are necessary.

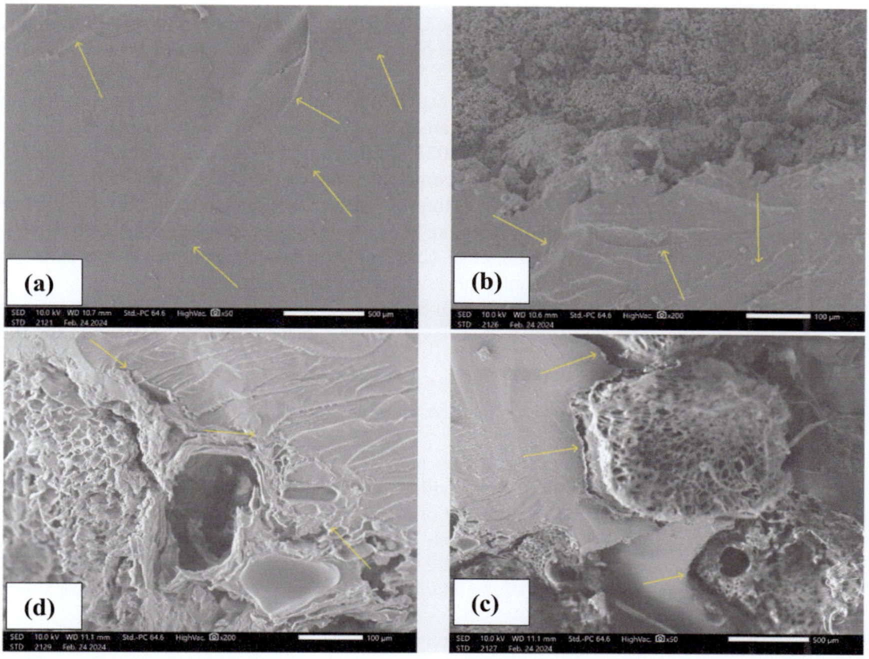

Fig. 4 SEM image of the **a** crack propagation and **b** fracture surface of the neat polyester. **c** Fracture surface and **d** interfacial adhesion between fiber and matrix of the P30 composite

References

1. Dimple, Singh GP, Mangal R (2023) A comprehensive review of natural fiber reinforced composite and their modern application. Mater Today Proc. 92: 542–548
2. Chung DDL (2019) A review of multifunctional polymer-matrix structural composites. Compos Part B Eng 160:644–660
3. García-Méndez RF, Cortés-Martínez CI, Almendárez-Camarillo A (2020) Thermochemical and Tensile Mechanical Properties of Fibers Mechanically Extracted from Leaves of Agave angustifolia Haw. J Nat Fibers 19:3171–3185
4. Benítez-Guerrero M, López-Beceiro J, Sánchez-Jiménez PE et al (2014) Comparison of thermal behavior of natural and hot-washed sisal fibers based on their main components: Cellulose, xylan and lignin. TG-FTIR analysis of volatile products. Thermochim Acta. 581: 70–86
5. Li M, Pu Y, Thomas VM, Yoo CG, Ozcan S, Deng Y, Ragauskas AJ (2020) Recent advancements of plant-based natural fiber–reinforced composites and their applications, Compos Part B: Eng 200:108254, ISSN 1359–8368, https://doi.org/10.1016/j.compositesb.2020.108254
6. Hariprasad K, Ravichandran K, Jayaseelan V et al (2020) Acoustic and mechanical characterisation of polypropylene composites reinforced by natural fibres for automotive applications. J Mater Res Technol 9:14029–14035
7. Gupta A (2020) Improvement of physiochemical properties of short bamboo fiber-reinforced composites using ceramic fillers. J Nat Fibers 17:1582–1593
8. Khalid MY, Al Rashid A, Arif ZU et al (2021) Natural fiber reinforced composites: Sustainable materials for emerging applications. Results Eng 11:100263
9. Smith N (2015) Euterpe oleracea. Geobot Stud, 239–258

10. Faria Viana L, Kingo Oyama Homma A, José Elias Amorim de Menezes A et al (2021) Análise econômica do cultivo de açaizeiro (Euterpe oleracea Mart.) Irrigado No Nordeste Paraense. Terceira Margem Amaz 2021; 7: 155–169
11. Lameira Nogueira O, Benito B, Calzavara G et al (1998) Manejo de açaizeiros (Euterpe oleracea Mart.) Cultivados em latossolo amarelo na amazônial. 53–59
12. Rahman A, Fehrenbach J, Ulven C et al (2021) Utilization of wheat-bran cellulosic fibers as reinforcement in bio-based polypropylene composite. Ind Crops Prod 172:114028
13. Anuar H, Zuraida A, Kovacs JG et al (2012) Improvement of mechanical properties of injection-molded polylactic acid-kenaf fiber biocomposite. J Thermoplast Compos Mater 25:153–164
14. Chirayil CJ, Mathew L, Hassan PA et al (2014) Rheological behaviour of nanocellulose reinforced unsaturated polyester nanocomposites. Int J Biol Macromol 69:274–281
15. Setter C, Silva FTM, Assis MR et al (2020) Slow pyrolysis of coffee husk briquettes: Characterization of the solid and liquid fractions. Fuel 261. https://doi.org/10.1016/j.fuel.2019.116420
16. Maradini G da S, Oliveira MP, Guanaes GM da S et al (2020) Characterization of polyester nanocomposites reinforced with conifer fiber cellulose nanocrystals. Polymers (Basel); 12: 1–19
17. Costa UO, Nascimento LFC, Garcia JM et al (2020) Evaluation of Izod impact and bend properties of epoxy composites reinforced with mallow fibers. J Mater Res Technol 9:373–382
18. Albdiry MT, Yousif BF (2019) Toughening of brittle polyester with functionalized halloysite nanocomposites. Compos Part B Eng 160:94–109
19. Goriparthi BK, Suman KNS, Mohan RN (2012) Effect of fiber surface treatments on mechanical and abrasive wear performance of polylactide/jute composites. Compos Part A Appl Sci Manuf 43:1800–1808
20. Thirumalaisamy R, Pavayee SS (2018) Investigation of physico-mechanical and moisture absorption characteristics of raw and alkali treated new agave angustifolia marginata (AAM) fiber. Medziagotyra 24:53–58

Compressive Properties of Epoxy Composites Reinforced with Sugarcane Bagasse Powder for High-Performance Flooring

Juliana Sousa Gomes Ribeiro, Victor Paes Dias Gonçalves, Noan Tonini Simonassi, Carlos Maurício Fontes Vieira, and Felipe Perissé Duarte Lopes

Abstract The growing demand for sustainable materials drives the valorization of agro-industrial residues. In this scenario, sugarcane bagasse, an abundant agro-industrial byproduct, emerges as a promising alternative for composite materials, promoting the circular economy. This research explores the potential of incorporating these residues into epoxy composites to develop sustainable, cost-effective, high-performance coatings. The study aims to contribute to the understanding of how natural fillers affect the mechanical properties of epoxy resins and provide insights into the benefits of using renewable resources in composite material production. This study investigates the hardness properties of epoxy composites reinforced with sugarcane bagasse powder; the reinforcement varies between 5 to 20 vol.%. A compression test will be conducted on the composite, in order to identify the force required for its rupture.

Keywords Hardness · Epoxy composites · Sugarcane bagasse power · Compression tests

Introduction

The global industry has increasingly sought to align with the sustainability policies established by the United Nations member states in the 2030 Agenda, which aims to address critical global challenges, including poverty, inequality, climate change, environmental degradation, peace, and justice. In total, 17 goals were defined, some of which are directly related to industrial activities and to the development of new materials that are eco-friendly and sustainable [1].

J. S. G. Ribeiro (✉) · V. P. D. Gonçalves · N. T. Simonassi · C. M. F. Vieira · F. P. D. Lopes
State University of the Northern Rio de Janeiro—UENF, Advanced Materials Laboratory—LAMAV, Campos Dos Goytacazes, Brazil
e-mail: ju.sgomes@hotmail.com

H. A. C. Lopera et al. (eds.), *Natural Fibers and Biocomposites 2026*, The Minerals, Metals & Materials Series, https://doi.org/10.1007/978-3-032-13577-3_3

The significant increase in the generation of harmful byproducts from industrial activities over recent decades has driven an equally strong effort toward their treatment. Initially, the focus was solely on reducing environmental and human health risks. Later, this purpose was broadened to include the reintegration of treated materials into the economic cycle, in line with the principles of sustainability and the circular economy [2, 3].

In this context, efforts are directed toward providing proper disposal for waste, while simultaneously seeking eco-friendly materials capable of replacing non-sustainable ones, such as petroleum-based products [4–6].

One solution widely investigated by researchers is the use of natural fibers in the production of materials that traditionally rely on synthetic fibers or petroleum-based, non-sustainable byproducts [4–7].

A composite material is the physical and chemical combination of two or more materials. These materials consist of a matrix and one or more dispersed phases, which are incorporated into the initial matrix [8, 9].

The dispersed phases can serve several purposes, among them reinforcement, where they act as an aggregate within the material and enhance its mechanical, chemical, or physical properties [9]. The origin of the matrix and the filler, as well as the shape and proportion of the reinforcement, directly affect the interface between the matrix and the dispersed phase, thereby automatically influencing the properties of the final composite material [4, 8].

Natural fibers offer economic, physical, and environmental advantages compared to synthetic fibers, exhibiting lower density, lower cost, and reduced environmental impact. Their use in composites tends to be beneficial from both an economic and ecological perspective. Furthermore, these renewable fibers can be utilized in the production of new sustainable products, reinforcing the concept of a circular economy. The biological, chemical, and environmental properties of natural fibers can significantly contribute to enhancing the final characteristics of composites, expanding their potential applications across various fields [10].

Lignocellulosic fibers have broad applications due to advantageous characteristics such as low cost, which in some cases can be nearly negligible, good stiffness, and reduced environmental impact [7]. Another important feature of lignocellulosic fibers is their low specific weight; glass fiber, a widely used synthetic fiber, has a density of 2.4 g/cm^3, whereas natural fibers have a density of 1.2–1.6 g/cm^3, making natural fibers ideal for composites that need to be lightweight. These attributes make them a promising and sustainable alternative for reinforcing composite materials [5, 7].

Different types of lignocellulosic fibers can be used, such as sisal, hemp, jute, sugarcane bagasse, and flax, and they can be short, long, or continuous, which directly influences the mechanical performance of the composite. The adhesion between the matrix and the fiber is crucial and can be optimized through chemical or physical surface treatments of the fibers, ensuring better load transfer and increased material durability [8].

Natural fiber-reinforced composites with polymer matrices have gained prominence in various fields due to their combination of lightness, adequate mechanical strength, and sustainability. In the automotive industry, these materials are used in door panels, interior trim panels, seats, and structural components, replacing synthetic fibers and contributing to weight reduction and lower CO_2 emissions [11].

In the construction sector, resin-based composites reinforced with natural fibers are used in structural panels, partitions, high-performance coatings, and thermal/acoustic insulation, offering durability, low weight, and reduced environmental impact compared to conventional materials [12].

Furthermore, these composites have applications in consumer and sports products, such as furniture, sports boards, and musical instruments, where the combination of aesthetics, lightness, and strength is highly valued. The ability to process natural fibers with different types of resins, such as epoxy or polyester, allows for tailoring mechanical and thermal properties according to the intended application [13].

Sugarcane bagasse is an abundant byproduct of the sugar-alcohol industry, composed mainly of lignocellulosic fibers [14].

The global production of sugarcane bagasse was approximately 2.03 billion tons in 2023, and Brazil is the largest producer in the world, reaching 713.214.1 thousand tons of sugarcane produced in the 2023/2024 harvest [15, 16].

According to Gurgel et al., in Brazil, approximately 30% of sugarcane bagasse is discarded, and for every ton of crushed sugarcane, 250 kg of bagasse is generated. In other words, based on the values presented, Brazil alone produced around 214 thousand tons of bagasse, of which 53.49 thousand tons were not utilized [17].

After juice extraction, the crushed sugarcane is used to produce bagasse, a fibrous substance composed of the plant stalk, including both the inner pith and the outer rind. The rind has a tough fibrous structure responsible for protecting the soft, spongy inner region, while the stalk contains long, fine fibers randomly distributed and bonded by hemicellulose and lignin. The inner region contains smaller fibers, with sucrose being the predominant component [14, 18].

Dried sugarcane bagasse is chemically composed mainly of cellulose [40–50%], hemicellulose [25–35%], and lignin [20–25%], with small amounts of ash and wax [14].

Cellulose provides rigidity and mechanical strength, while hemicellulose and lignin contribute to structural cohesion and thermal stability of the material. The fibers extracted from bagasse have a low specific weight, ranging from 1.2 to 1.5 g/cm^3, making them ideal for composites that require lightness without compromising mechanical strength. Additionally, these fibers are renewable, biodegradable, and low cost, enhancing their environmental and economic appeal [14, 18].

The purpose of this study is to analyze the feasibility of a composite material system combining epoxy and natural fiber as an alternative to materials containing non-biodegradable components that are harmful to the environment, with applications in high-performance coatings.

Materials and Methodology

The present study aims to analyze the compressive properties of epoxy composites reinforced with sugarcane bagasse fiber. For the preparation of the mixture and analysis, the following steps were followed and Fig. 1 schematically illustrates the steps described above:

1. Sugarcane bagasse was obtained from local sugarcane juice vendors.
2. The material was dried at room temperature for 48 h and then placed in an oven at 60 °C for an additional 48 h.
3. The dried bagasse was cut and ground using a knife mill.
4. The resulting powder was sieved through a 42-mesh stainless steel screen.
5. Composites were prepared using the sieved powder in proportions of 5%, 10%, 15%, and 20% by mass of sugarcane bagasse, corresponding to 95%, 90%, 85%, and 80% by mass of the resin–hardener system [DGEBA 2050/DETA 3131—AVIPOL], respectively.
6. Test samples were produced in open silicone molds with dimensions of 22.5 × 22.5 mm, according to ASTM C579–23 [19].
7. Curing was carried out at room temperature [25 °C] for 24 h.
8. Compression tests were conducted using an Instron 5582 universal testing machine at the LAMAV/UENF laboratory, with a testing speed of 1.3 mm/min.
9. Data analysis was performed using Origin software, from which the analytical graphs were generated.
10. Microstructural Analysis by Scanning Electron Microscopy (SEM) TESCAN MIRA4.

Fig. 1 Schematic of sample preparation and compression testing

Ten specimens of each formulation were prepared and tested, as recommended by ASTM C579 [19].

Results and Discussion

As a result of the conducted test, the graph shown in Fig. 2 was developed, illustrating compressive strength in MPa on the y-axis and the tested specimen on the x-axis, where the values represent the mass percentage of residue in the sample and the values shown correspond to Table 1.

The presented values can be analyzed from two perspectives: considering the standard deviation or the compressive strength. Evaluating the standard deviation, the 15% formulation shows the best result, whereas, considering only compressive strength, the 10% formulation is the most suitable.

However, test data should not be analyzed without an appropriate and validated parameter. In Brazil, high-performance coatings follow the ABNT NBR 14,050/

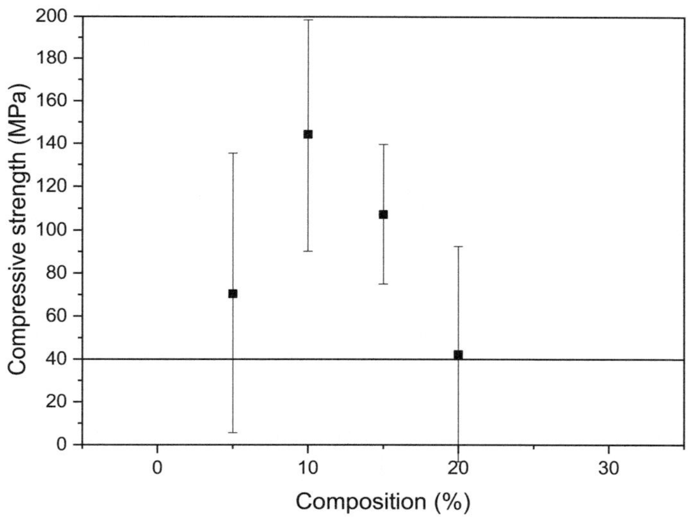

Fig. 2 Compressive strength test results

Table 1 Compressive strength test data

Composition (%)	Compressive strength [MPA]
5	70,44 (±64,88)
10	144,32 (±51,16)
15	107,28 (±32,238)
20	42,29 (±50,23)

1998, which establishes minimum parameters for high-performance coatings. There is no international standard specifying this material. Therefore, the initial comparison of the studied material will be made with the NBR. The standard presents Table 2, which has been adapted [20].

Analyzing Table 2 and the data obtained from the test, it is possible to state that the studied material meets at least the requirements for types 2, 3, 4, and 5 with the 20% formulation. The composite with the second-lowest compressive strength exceeds the minimum standard requirement by 76%, corresponding to the 5% bagasse composite. In other words, from a 5% addition of sugarcane bagasse by mass to 95% of the epoxy system [DGEBA 2050/DETA 3131—AVIPOL], a composite is produced that can be used as a high-performance coating according to the Brazilian standard. Evaluating the other compositions, some formulations surpass the minimum requirement by more than 200%.

Despite this high performance, increasing the reinforcement content to 20% resulted in a sharp decrease in compressive strength [42.29 MPa]. Although still compatible with the standard requirements, this indicates the existence of a critical limit for fiber incorporation. This behavior may be related to poor dispersion of the bagasse powder within the matrix, the formation of agglomerates, and reduced interfacial adhesion between fiber and resin. Additionally, potential defects during manual specimen fabrication may have contributed.

Similar phenomena were reported by Annamalai et al. in sisal and abaca composites, and by in pineapple fiber-reinforced systems, where higher reinforcement contents compromised load transfer [21–23].

Another relevant aspect is the magnitude of the observed standard deviations, which, in some cases, were high. This behavior suggests heterogeneity in the distribution of the reinforcement during molding, indicating that the sample preparation process requires improvement to ensure greater reproducibility. Recent studies highlight the importance of chemical treatments on fibers to enhance the fiber–matrix interface, reduce variations between samples, and increase the stability of the composite under different processing conditions [7].

Table 2 Compressive strength requirements according to NBR ABNT 14,050/1998 [20]

Type	Class	Description	Minimum compressive strength [MPa]
01	Monolithic	Trowel-applied	45
02		Self-leveling	40
03		Multiple layers	40
04	Coating	Low thickness	N.A.
		High thickness	
05	Decorative monolithic	Trowel-applied	= 1, 2 or 3
		Self-leveling	
		Multiple layers	

Compared to the literature, the performance obtained in this study is competitive or even superior to other polymeric systems reinforced with natural fibers. Gowda et al. reported a compressive strength of 45 MPa in jute–polyester composites, while Senthilkumar et al. observed modest improvements in composites with pineapple leaf fibers. The results obtained here demonstrate that epoxy composites with sugarcane bagasse can achieve significantly higher values, highlighting their potential as a high-performance alternative for industrial flooring and coatings subjected to mechanical stresses [21, 22].

Figure 3a–d show Scanning Electron Microscopy (SEM) micrographs obtained using the TESCAN MIRA4 microscope available at LAMAV/UENF. Images (a) and (b) correspond to the composite containing 5 wt% sugarcane bagasse fiber, whereas images (c) and (d) refer to the 20 wt% fiber-reinforced sample.

The micrographs shown in Fig. 3a and b clearly reveal dispersed particulate regions within the epoxy matrix. A good fiber–matrix integration is observed, with adequate interfacial adhesion, although the matrix phase predominates. The particulates appear spatially separated, contributing to moderate porosity in the composite. However, based on the compressive strength results, this porosity does not significantly affect the mechanical integrity of the material.

In contrast, Fig. 3c and 3d depict a composite in which the distinction between fiber and matrix becomes less evident, indicating a more homogeneous distribution of the

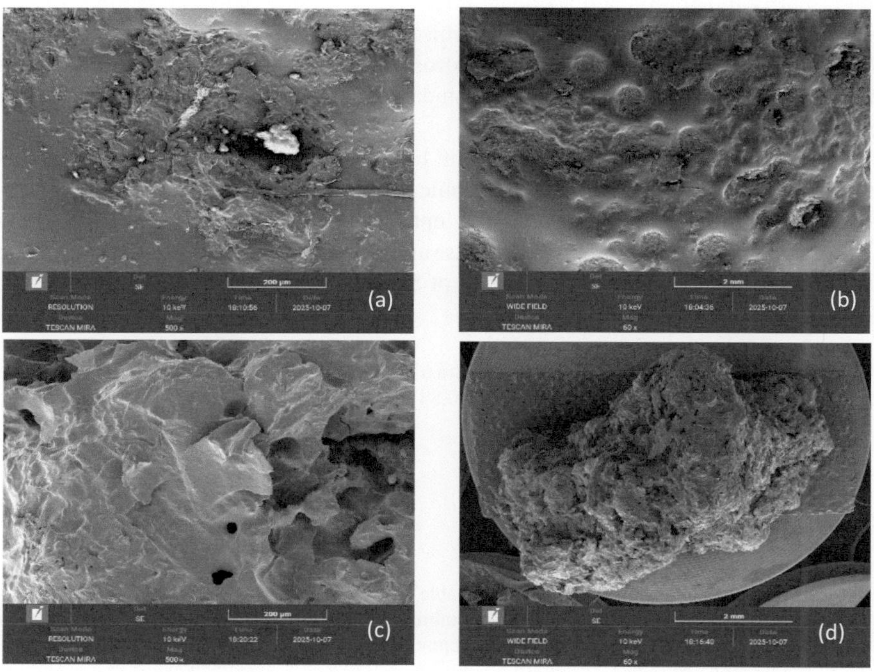

Fig. 3 Microstructural analysis by scanning electron microscopy (SEM)

phases. Nevertheless, larger and more frequent pores can be observed throughout the structure, suggesting void coalescence and poorer dispersion. The absence of clearly defined particulates implies agglomeration, which, in light of the compressive data, suggests that increased porosity leads to reduced mechanical strength.

Conclusion

Considering the presented study, the proposed composites are viable for their intended application according to the reference standard. Additionally, it can be observed that other researchers have also reported results supporting the improvement of properties through the introduction of fibers into polymeric matrices.

The results demonstrated that incorporating sugarcane bagasse powder into the epoxy matrix significantly affects the compressive behavior of the composites. The highest compressive strength was achieved for the 10 wt% formulation, reaching 144.32 MPa, which represents an increase of approximately 105% compared to the 5 wt% sample. Even the 20 wt% composite exhibited 42.29 MPa, exceeding the minimum requirement of 40 MPa established by ABNT NBR 14,050/1998 for high-performance coatings, confirming the technical feasibility of the proposed material.

SEM microstructural analysis revealed that the 5 wt% composite exhibited good fiber–matrix integration and adequate interfacial adhesion, with moderately spaced pores that did not significantly compromise compressive strength. In contrast, the 20 wt% composite showed increased porosity, particle agglomeration, and poorer dispersion, resulting in reduced mechanical performance, in agreement with the compressive data.

Therefore, the controlled addition of lignocellulosic fibers to epoxy systems enhances both sustainability and mechanical performance, provided that the reinforcement content remains within the optimal range (10–15 wt%). This study confirms the potential of sugarcane bagasse as a renewable filler for high-performance composites and highlights the relevance of SEM analysis in correlating microstructural features with mechanical behavior.

Acknowledgements The authors thank the Brazilian agencies CNPq, process number 302976/2022-1, and FAPERJ, process number E-26/200.847/2021.

References

1. dpicampaigns. Take Action for the Sustainable Development Goals [Internet]. United Nations Sustainable Development. [citado 4 de setembro de 2025]. Disponível em: https://www.un.org/sustainabledevelopment/sustainable-development-goals/
2. Korniejenko K, Mikuła J, Brudny K, Aruova L, Zhakanov A, Jexembayeva A et al (2025) A review of industrial by-product utilization and future pathways of circular economy: geopolymers as modern materials for sustainable building. Sustainability. janeiro de 17(10):4536

3. Wibisono DAS, Saw CY, Wu TY, Chau CF (2025) Advancing industrial food byproduct management: strategies, technologies, and progress in waste reduction. Processes. janeiro de 13(1):84
4. Awed RMA, Karahan M, Ahmed HAM, Gismallah U (2022) A review on natural fibers' for biopolymer composites. trends text eng fash technol. 7 de setembro de. 7(2):820–35
5. Karimah A, Ridho MR, Munawar SS, Ismadi, Amin Y, Damayanti R, et al (2021) A comprehensive review on natural fibers: technological and socio-economical aspects. Polymers. janeiro de. 13(24):4280
6. Kamarudin SH, Mohd Basri MS, Rayung M, Abu F, Ahmad S, Norizan MN, et al (2022) A Review on natural fiber reinforced polymer composites (NFRPC) for sustainable industrial applications. Polymers. janeiro de. 14(17):3698
7. Mahmud MdZA, Rabbi SMF, Islam MdD, Hossain N. Synthesis and applications of natural fiber-reinforced epoxy composites: A comprehensive review. SPE Polym. 2025;6(1):e10161
8. Hsissou R, Seghiri R, Benzekri Z, Hilali M, Rafik M, Elharfi A. Polymer composite materials: A comprehensive review. Compos Struct. 15 de abril de 2021;262:113640.
9. Farahmand B (2025) Understanding of composite materials (constituents of composites). In: Farahmand B (ed) Fundamentals of composites and their methods of fabrications: PMCs, MMCs, and CMCs. Springer Nature Switzerland, pp 1–17. https://doi.org/10.1007/978-3-031-80201-0_1
10. Challenges and advancement in water absorption of natural fiber-reinforced polymer composites. Polym Test. 1° de julho de 2023;124:108083
11. Azman MA, Asyraf MRM, Khalina A, Petrů M, Ruzaidi CM, Sapuan SM, et al (2021) Natural fiber reinforced composite material for product design: a short review. polymers. 9 de junho de. 13(12):1917
12. Ntsie OD, Phiri R, Boonyasopon P, Rangappa SM, Siengchin S (2025) Advancing sustainable infrastructure: natural fiber-reinforced composites in engineering. Discov Appl Sci. 4 de agosto de. 7(8):884
13. Ichim M, Muresan EI, Codau E (2024) Natural-fiber-reinforced polymer composites for furniture applications. Polymers. janeiro de 16(22):3113
14. Mahmud MdA, Anannya FR (2021) Sugarcane bagasse—A source of cellulosic fiber for diverse applications. Heliyon. 13 de agosto de. 7(8):e07771
15. FAOSTAT [Internet]. [citado 4 de setembro de 2025]. Disponível em: https://www.fao.org/faostat/en/#data/QCL/visualize
16. boletim-cana-de-acucar-4o-levantamento-2024–25 [Internet]. [citado 4 de setembro de 2025]. Disponível em: https://www.gov.br/conab/pt-br/atuacao/informacoes-agropecuarias/safras/safra-de-cana-de-acucar/arquivos-boletins/4o-levantamento-safra-2024-25/boletim-cana-de-acucar-4o-levantamento-2024-25
17. PL-Gerenciamento-tratamento-2024.pdf [Internet]. [citado 4 de setembro de 2025]. Disponível em: https://www.alice.cnptia.embrapa.br/alice/bitstream/doc/1167011/1/PL-Gerenciamento-tratamento-2024.pdf
18. Zafeer MohdK, Prabhu R, Rao S, Mahesha G, Bhat KS (2023) Mechanical characteristics of sugarcane bagasse fibre reinforced polymer composites: a review. Cogent Eng. 31 de dezembro de. 10(1):2200903
19. D01 Committee. Test Methods for Compressive Strength of Chemical-Resistant Mortars, Grouts, Monolithic Surfacings, and Polymer Concretes [Internet]. ASTM International; [citado 4 de setembro de 2025]. Disponível em: http://www.astm.org/cgi-bin/resolver.cgi?C579-23
20. NBR 14050.1998—Sistemas de revestimentos de alto desempenh

21. Senthilkumar K, Saba N, Chandrasekar M, Jawaid M, Rajini N, Alothman OY et al (2019) Evaluation of mechanical and free vibration properties of the pineapple leaf fibre reinforced polyester composites. Constr Build Mater. janeiro de 195:423–431
22. Munikenche Gowda T, Naidu ACB, Chhaya R (1999) Some mechanical properties of untreated jute fabric-reinforced polyester composites. Compos Part Appl Sci Manuf. 1° de março de. 30(3):277–84
23. Annamalai K, Soundararajan S, Kalidas S, Marialueedass N (2024) Explorations into the mechanical properties of composites reinforced with sisal and abaca natural fibers. Matér Rio Jan. 29:e20240585

Part II
Sustainable Structural Composites and Their Applications

Characterization of Polyamide PA11 from the Reuse of Oil and Gas Industry Pipes

Jhon Peter Aleixo Ferreira Junior, Victor Dias Gonçalves, Noan Tonini Simonassi, and Felipe Perisse Duarte Lopes

Abstract One of the major problems generated by operations on platforms is the disposal of polymeric materials used to coat decommissioned pipelines—polyamide PA11—which, when disposed of incorrectly, can pose many risks to the environment, making it necessary to look for ways of reusing them. The aim of this research is to evaluate the mechanical, physical and chemical parameters of polyamide PA11 from the reuse of pipes from the oil and gas industry through hardness tests (Shore D), traction, density, infrared spectroscopy (FTIR), wettability, dilatometry, and DSC/TGA. The initial results indicate that the material is capable of presenting good properties even after reuse cycles. The conclusion reached is that the properties of polyamide PA11 are very good, especially for pipelines in the oil and gas industry, and that its reuse would be essential not only from an environmental point of view, but also from an economic one.

Keyword Decommissioning · Polymer · Sustainability

Introduction

The decommissioning of pipelines in the oil and gas industries represents a very important step in safely and sustainably ending oil and gas production operations [1]. This process includes cleaning facilities, plugging wells, disconnecting subsea systems, and dismantling structures, with a primary focus on reducing environmental impacts and applying sustainable practices [1, 2]. One of the main challenges is the disposal of polymeric materials used as pipe coatings, such as polyamide PA11. If disposed of improperly, these materials can pose risks to the environment, making it necessary to investigate possibilities for reuse [1, 2].

Polyamide PA11 is widely used as an internal coating for flexible pipes, mainly due to its chemical resistance to hydrocarbons and its good mechanical properties,

J. P. A. F. Junior (✉) · V. D. Gonçalves · N. T. Simonassi · F. P. D. Lopes
Universidade Estadual do Norte Fluminense—UENF, Rua Professora Julia Boynard, Campos Dos Goytacazes, Brazil
e-mail: jhon.peter.aleixo@gmail.com

H. A. C. Lopera et al. (eds.), *Natural Fibers and Biocomposites 2026*, The Minerals, Metals & Materials Series, https://doi.org/10.1007/978-3-032-13577-3_4

being one of the most widely used materials in the offshore industry, where it emerged as one of the first materials for the manufacture of flexible pipes [3, 4]. As it is a thermoplastic material, it can be easily processed via extrusion into pipes of different diameters, thicknesses, and lengths through a continuous process [4]. One of the main challenges in using thermoplastic materials, such as PA11, is related to predicting their long-term behavior under severe conditions, such as those found in offshore operations [3].

The aim is to show whether the material in question still has the capacity and properties to perform its function, such as coating platform pipes even after reuse, contributing to a more sustainable transition in the oil and gas industry. The objective of the research is to evaluate the mechanical, physical, and chemical parameters of PA11 polyamide from the reuse of pipes from the oil and gas industry through hardness (Shore D), density, infrared spectroscopy (FTIR), wettability, dilatometry, TGA/DSC, and tensile tests.

Materials and Methods

Basically, the polymer (PA11) used in this research comes from decommissioned pipes supplied by a company in the Port of Açu (São João da Barra, RJ). To prepare the samples, the material was cut using a band saw, allowing us to prepare the polymer in the correct size, in accordance with the standards for the respective tests (Fig. 1).

Fig. 1 Polyamide PA 11

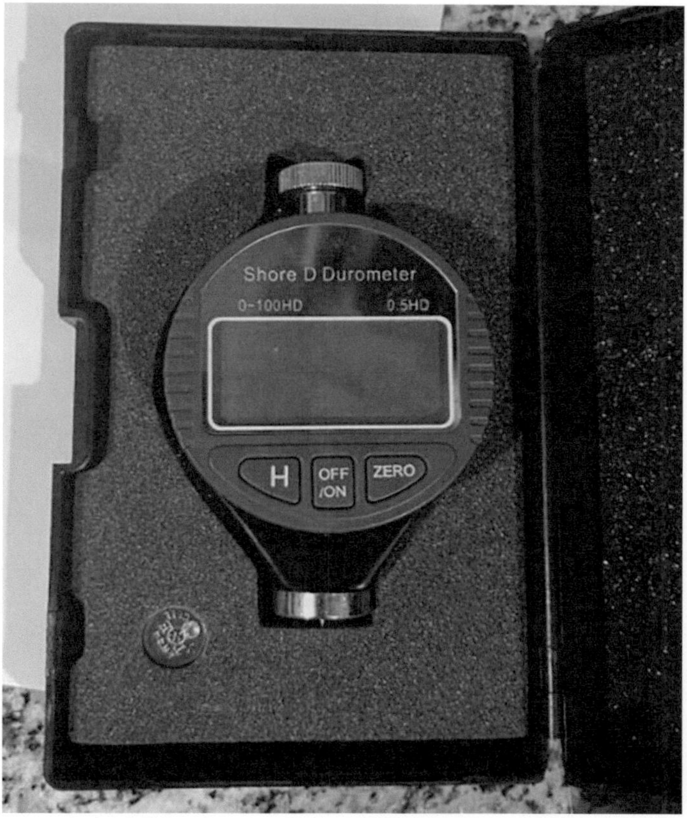

Fig. 2 Shore D hardness tester

Hardness Test

The hardness test was performed using a Shore D hardness tester (Fig. 2), an instrument used to perform measurements, which consists of a hardened steel cone indenter that applies force to the material and measures the depth of the indentation. The Shore D scale is ideal for harder and more rigid materials such as polyamide (PA11). The sample was prepared in accordance with standard ASTM D2240-15 [10].

Density Test

The density test was performed using pycnometry, a model that measures the buoyancy force exerted on a body immersed in a fluid, in accordance with ASTM D792-20.

Small pieces of polyamide were cut manually, totaling a mass of 1 (one) gram of material (Fig. 3).

The test consisted of the following steps:

1. Weighing the empty pycnometer.
2. Weighing the pycnometer with the material inside.
3. Adding water to the pycnometer, ensuring there are no bubbles.
4. Weighing the pycnometer with the material and water.
5. Calculating the density.

Fig. 3 Picnometer

$$D_r = \frac{B - A}{(D - A) - (C - B)}$$

where:

A: Empty pycnometer
B: Pycnometer + Material.
C: Pycnometer + Material + Water.
D: Pycnometer + Water.

FTIR Test

The FTIR test was performed using an IRTracer-100 infrared spectrometer manufactured by Shimadzu. For this test, after cutting the samples to obtain the necessary measurements, they were mainly kept in the place where they were cut in order to avoid any interference in the spectrometry results (Fig. 4).

Fig. 4 FTIR machine

Wettability Test

In order to determine the affinity of the polymer in question with water, a wettability test was performed. The test was carried out in accordance with ASTM D7334-08.

The steps basically consisted of:

1. Preparation of the sample, ensuring a smooth, dirt-free surface.
2. Applying a drop of water to the surface of the material.
3. Measuring the contact angle.
4. Analyzing the results.

Dilatometry Test

The thermal expansion behavior of polyamide PA11 was evaluated using a dilatometry test. The test was carried out in accordance with ASTM E228-22. The samples were placed on the dilatometer support and subjected to a controlled heating program under an inert atmosphere to prevent oxidation processes. During the test, the linear dimensional variation ($\Delta L/L$) of the sample as a function of temperature was monitored using a displacement transducer coupled to the equipment. From the dilatometric curve obtained, it was possible to identify parameters such as the coefficient of linear thermal expansion (CTE), as well as the characteristic thermal transitions of the polymer, such as the glass transition (Tg) and crystalline melting (Tm).

TGA/DSC Testing

The thermal stability and thermal transitions of polyamide PA11 were evaluated using thermogravimetric analysis (TGA) and differential scanning calorimetry (DSC). The test was carried out in accordance with ASTM. In the TGA tests, the samples were subjected to controlled heating in a nitrogen atmosphere at a predefined heating rate, recording the variation in mass as a function of temperature. This procedure allowed the onset of thermal degradation and the stability range of the material to be identified. For DSC analysis, the samples were subjected to heating and cooling cycles, also under an inert atmosphere, and the variations in heat flow were monitored.

Tensile Testing

The mechanical properties of polyamide PA11 were determined by means of tensile tests, performed in accordance with standard ASTM C307-23. The test specimens were made in a standardized format and subjected to uniaxial loading in a universal testing machine equipped with a load cell appropriate for the material's strength range. The test was conducted at a controlled deformation speed (50 mm/min), continuously recording the variation in the force applied and the elongation of the sample until rupture. From the stress–strain curve obtained, characteristic parameters of the mechanical behavior of PA11 were determined, such as modulus of elasticity, yield strength, maximum tensile strength, and elongation at break.

Results and Discussion

The Shore D hardness test revealed contrasting results between virgin PA11 (Arkema) and PA11 obtained from pipelines already used in the industry. The virgin material supplied by Arkema presented a hardness of approximately 71 Shore D, while the polymer previously in service showed a value of only 49 Shore D. This difference of 22 units highlights a significant loss of surface rigidity after exposure to operational conditions.

The reduction in hardness can be associated with phenomena such as thermal and oxidative aging, interaction with hydrocarbons and aggressive fluids present in the industrial environment, as well as polymer chain degradation through scission or oxidation. These processes promote a decrease in crystallinity and molecular relaxation, resulting in a material that is less rigid and more susceptible to permanent plastic deformation at the surface.

From an application standpoint, Shore D hardness is directly related to resistance to wear, abrasion, and permanent surface deformation. Therefore, the decrease from 71 to 49D suggests that used PA11 may exhibit inferior performance in critical applications, such as pipeline coatings in the oil and gas industry, compromising its durability in service.

Thus, the results demonstrate that although the reuse of PA11 represents a sustainable alternative, the significant reduction in hardness observed must be carefully considered. A thorough evaluation regarding compliance with technical standards and safety requirements is essential before its reapplication (Fig. 5).

The density of polyamide PA11 obtained through pycnometry testing was 1.0 g/cm^3 on average, which makes it relatively light for many applications. However, when compared to polypropylene (PP), based on the literature, we can state that PA11 is denser than the other polymers in question 0.

The obtained density of 1.0 g/cm^3 is consistent with the typical density range for PA11 reported in the literature, which is between 1.03 and 1.05 g/cm^3 [4, 6]. This value is significantly higher than that of polypropylene (PP, ~0.9 g/cm^3), as mentioned in the literature [5, 6]. The maintained density close to the virgin material suggests that the bulk structure and degree of crystallinity might not have been drastically altered by the service life, which is a positive indicator for its potential reuse (Fig. 6).

Based on this wettability test shown in Fig. 7, we can state that polyamide PA11, with an average contact angle of 62.46, being a hydrophilic polymer, water wets the surface well. One of the reasons for this affinity is the presence of polar amide

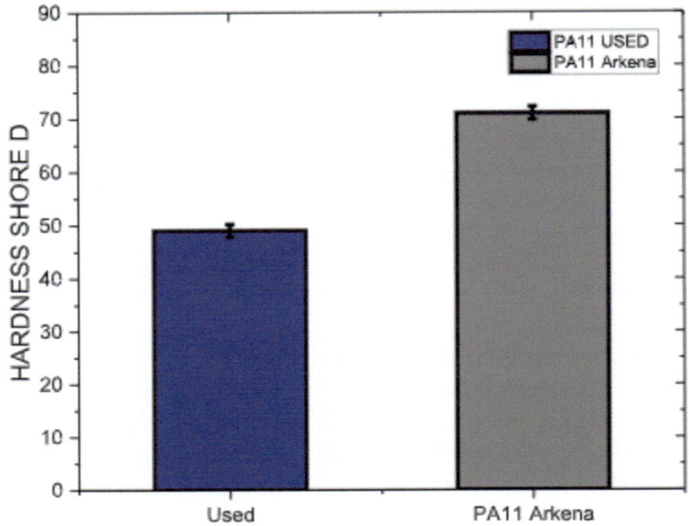

Fig. 5 Shore D hardness result

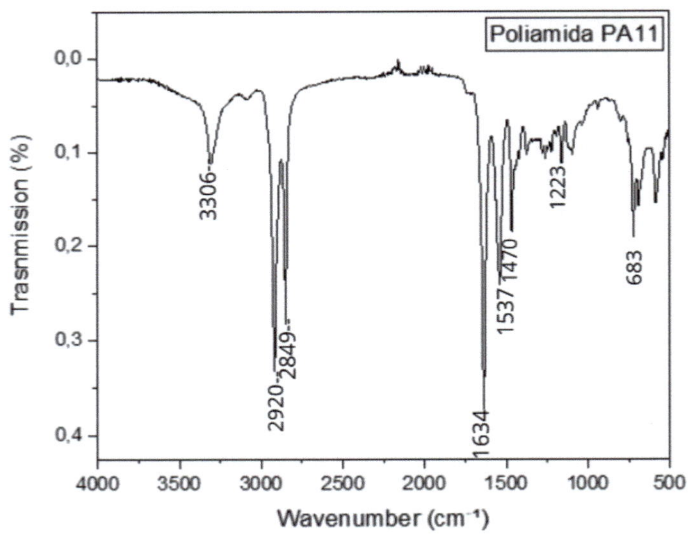

Fig. 6 FTIR result

groups (-CONH-) groups shown in Table 1 of the FTIR test, which contributes to this greater affinity with water. [8] In contrast, polymers that are also used with this affinity, such as polypropylene (PP) are hydrophobic, i.e., they have contact angles greater than 90°. This occurs because, unlike PA11, their chains are nonpolar and do not interact well with water. This characteristic is essential for application in industrial pipe coatings, as it favors applications in which adhesion and interaction with liquids are important.

The obtained contact angle of 62.46° is consistent with values reported for virgin PA11 in the literature, which typically classify it as a hydrophilic polymer due to the polar amide groups in its backbone [6]. This value is fundamentally different from that of hydrophobic polymers like polypropylene (PP) [5, 6]. This preserved hydrophilicity indicates that the surface chemistry, regarding its polarity, was largely maintained during its service life.

The dilatometric curve of polyamide PA11 (Fig. 8) shows the relative variation in length ($\Delta L/L_0$) as a function of temperature. A progressive increase in linear expansion is observed from approximately 25 °C, indicating the expected behavior of the polymer when heated. Between 40 and 70 °C, there is a change in the slope of

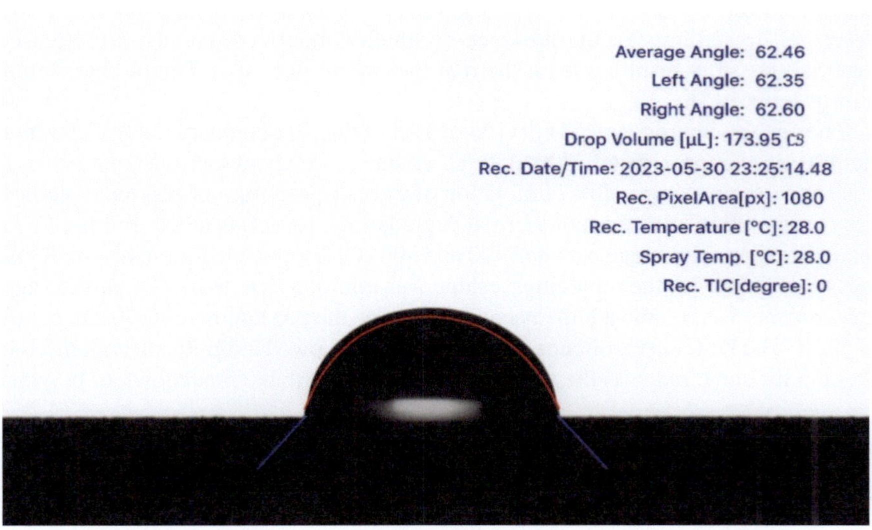

Fig. 7 Result wettability

Table 1 Position of bands and chemical bonds [4]

3306	N–H stretching
2920	CH2 asymmetric stretching
2849	CH2 symmetric stretching
1634	C = O stretching (amide I)
1537	C–N stretching and C = O in plane bending
1470	C = O and N-vicinal CH2 bending
1223	C–N stretching
683	CONH out-of-plane deformation (amide V)

the curve, attributed to the glass transition (Tg), which is associated with increased mobility of the polymer chains in the amorphous phase. As heating continues, the curve maintains a sharp increase up to about 140–150 °C, at which temperature an inflection occurs followed by a decrease in the value of $\Delta L/L_0$. This behavior is characteristic of the melting of the crystalline phase (Tm) of PA11, at which point structural rearrangement and loss of dimensional stability of the material occur. The peak observed in this region confirms the semi-crystalline nature of PA11 and its sensitivity to thermal transitions. Therefore, the dilatometry test clearly identified the relevant thermal transitions of PA11, highlighting both Tg and Tm, in addition to providing information on the dimensional stability of the polymer in relation to temperature, a crucial aspect for applications in conditions of thermal variation. The identified Tg and Tm ranges are in good agreement with values reported in the literature for PA11, where Tg is typically around 40–50 °C and Tm is between 180–190 °C for the virgin polymer [4, 6]. The Tm observed in this test appears lower than the typical 185 °C for virgin PA11. This shift could indicate a lower degree of crystallinity or the presence of smaller/imperfect crystals in the reused material, possibly resulting from thermal history or molecular weight degradation during service [4].

Thermogravimetric analysis (TGA) of PA11 (Fig. 9) demonstrated high thermal stability up to approximately 300–350 °C, a range in which no significant mass loss is observed, except for possible elimination of traces of moisture or residual volatiles. From 380–420 °C, the onset of thermal degradation (Tonset) is noted, followed by a main decomposition stage between 400 and 500 °C, responsible for massive material loss. Above 600 °C, the remaining residue was minimal (less than 3%), indicating a low content of inorganic additives and confirming the predominantly organic nature of PA11. The DSC curve, obtained simultaneously, showed a significant endothermic event in the same range as the mass loss, associated with the breakdown of polymer

Fig. 8 Dilatometry result

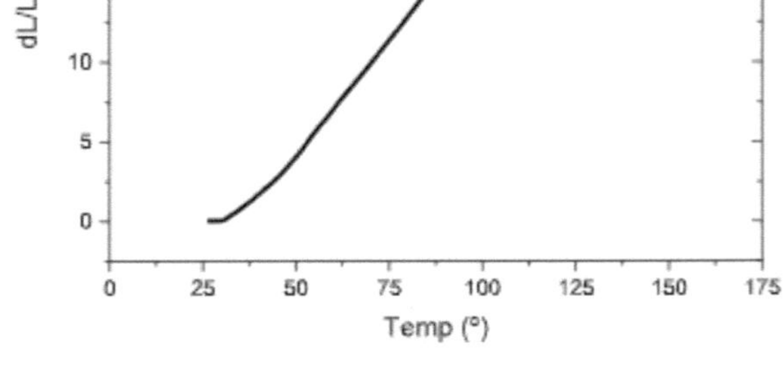

chains and the release of volatile products. After the main decomposition, the calorimetric signal stabilizes, in agreement with the behavior of the TGA curve. The presence of a single degradation stage shows that PA11 undergoes a single-stage decomposition process under an inert atmosphere. These results confirm that PA11 can be safely processed in the usual molding range (200–240 °C), well below the degradation onset temperature. In addition, the low solid residue fraction suggests a formulation with minimal mineral filler, which may contribute to good mechanical properties and structural stability, although it implies lower combustion resistance. This degradation profile is consistent with findings for virgin PA11 reported in the literature. Tey et al. [4] reported a single-stage decomposition for PA11 under nitrogen. The high degradation onset confirms that PA11 can be safely processed and used at temperatures well below 300 °C. The lack of significant residue is typical for unfilled or lightly stabilized PA11 and is advantageous for recycling processes but implies lower flame retardancy.

Figure 10 shows the stress–strain curve obtained from the tensile test of the PA11 polymer. The material exhibits the typical behavior of semicrystalline thermoplastic polymers, characterized by a relatively short elastic region followed by an extensive plastic deformation stage before fracture. In the initial stage of the test, up to approximately 10% strain, it is possible to identify the elastic region, where stress increases almost linearly with deformation. Beyond this point, the material enters the plastic regime, maintaining its ability to withstand load while undergoing continuous deformation, which demonstrates the ductile nature of PA11.The maximum tensile strength recorded was approximately 37–38 MPa, which represents the ultimate tensile strength of the material. This result is consistent with values reported

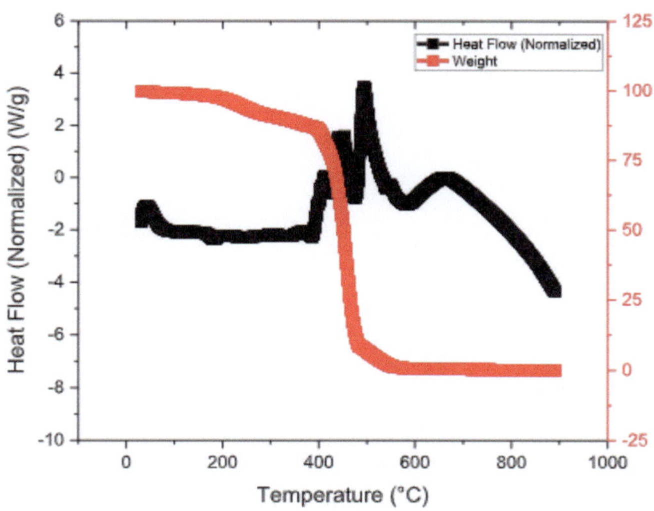

Fig. 9 Result TGA/DSC

in the literature and manufacturer data sheets, which typically range between 35–50 MPa, depending on processing conditions and the polymer's degree of crystallinity. Regarding elongation at break, PA11 exhibited a strain close to 100%, indicating that the material nearly doubled its original length before failure. This high deformability confirms the ductile character of the polymer, which is particularly advantageous in applications requiring both strength and the ability to absorb mechanical stresses, such as in the oil and gas sector, where PA11 is widely employed in flexible pipes and protective coatings. Therefore, the results reinforce the suitability of PA11 for applications demanding a combination of good mechanical strength and high ductility, which are essential properties for extending the service life of components subjected to mechanical and environmental stresses. Comparing these values with literature data for virgin PA11 provides critical insight. Virgin PA11 typically exhibits higher tensile strength, often in the range of 45–55 MPa, and very high elongation at break, often exceeding 200–300% [4, 6]. The obtained tensile strength of 37–38 MPa is at the lower end of the expected range, and the elongation at break of ~ 100% is significantly reduced compared to virgin material. This reduction in both ultimate strength and, more markedly, in ductility is a clear indicator of material degradation during its service life. This could be attributed to chain scission, leading to a reduction in average molecular weight [3, 4]. Despite this degradation, the retained mechanical properties are still substantial for many non-structural or semi-structural applications.

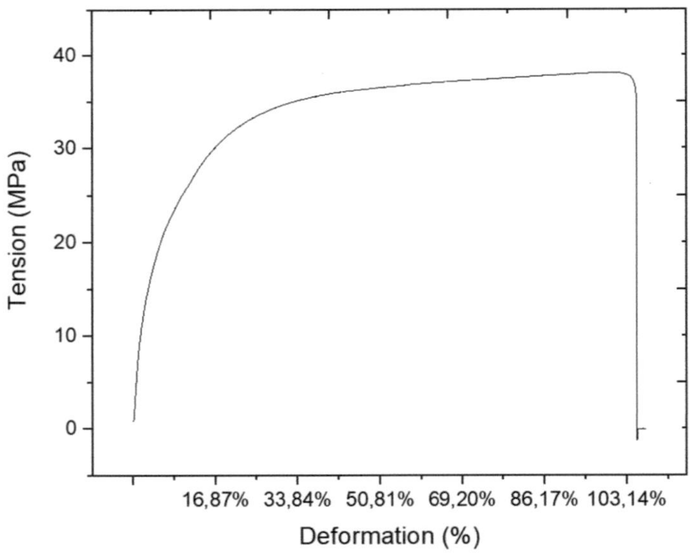

Fig. 10 Result tensile

Conclusion

The parameters observed in the study regarding polyamide PA11 reinforce the importance of conducting technical studies on the decommissioning process, seeking to find solutions aligned mainly with environmental sustainability. The reuse of polymers such as polyamide PA11 can reduce waste generation and add value to materials that would initially be discarded, thus contributing to more sustainable action by the oil and gas industry, not only from an environmental point of view but also from an economic one.

References

1. Radu I-C, Vadureanu A-M, Cozorici D-E, Blanzeanu E, Zaharia C (2025) Advancing Sustainability in Modern Polymer Processing: Strategies for Waste Resource Recovery and Circular Economy Integration. Polymers 17:522. https://doi.org/10.3390/polym17040522
2. Beena Unni A, Muringayil Joseph T (2024) Enhancing Polymer Sustainability: Eco-Conscious Strategies. Polymers 16:1769. https://doi.org/10.3390/polym16131769
3. Chollet C, Do AT (2013) Qualification of new polyamide for flexible flowlines and risers at elevated pressure and temperature. Presented at: offshore technology conference. Houston, Texas, USA. https://doi.org/10.4043/24019-MS
4. Tey WS, Cai C, Zhou K (2021) A comprehensive investigation on 3D printing of polyamide 11 and thermoplastic polyurethane via multi jet fusion. Polymers. 13(2139). https://doi.org/10.3390/polym13132139
5. Coutinho FM, Mello IL, Santa Maria LC (2011) Polyolefins, polyamides and polystyrene as matrix materials in wood plastic composites: A comparative study. Polymer 52(13):3014–3020. https://doi.org/10.1016/j.polymer.2011.02.002
6. Mark JE editor (2009) Polymer data handbook. 2nd ed. New York: Oxford University Press. 2009. p 1030. Minaspol Plásticos. Polipropileno. Minaspol. Disponível em: https://minaspol.com.br/produtos/polipropileno/
7. ASTM D2240–15; Standard test method for rubber property—durometer hardness. American Society for Testing Materials: West Conshohocken, PA, USA, 2015
8. ASTM D792–20; Standard Test Methods for Density and Specific Gravity (Relative Density) of Plastics by Displacement. ASTM International: West Conshohocken, PA, USA, 2020
9. ASTM D7334–08; Standard Practice for Surface Wettability of Coatings, Substrates and Pigments by Advancing Contact Angle Measurement. ASTM International: West Conshohocken, PA, USA, 2008
10. ASTM E228–22: Standard Test Method for Linear Thermal Expansion of Solid Materials with a Push-Rod Dilatometer. West Conshohocken, PA: ASTM International, 2022. https://doi.org/10.1520/E0228-22

Part III
Manufacturing and Innovative Sustainable Composites and Biocomposites

Alkali-Treated Jute Fibre Biocomposites for Lightweight and Eco-Friendly Engineering Applications

Dhaneshwar Prasad Sahu, Ramyaranjan Das, Jagesh Kumar Prusty, and Sukesh Chandra Mohanty

Abstract This study investigates the enhancement of mechanical properties in eco-friendly and biodegradable jute fibre-reinforced epoxy composites through alkali surface modification. Jute fibers (*Corchorus capsularis*) were used as reinforcement and epoxy L-12 as the matrix. To improve interfacial bonding and structural performance, the raw jute fibres underwent dewaxing followed by alkaline treatment using potassium hydroxide (KOH) at varying concentrations (2, 4, 6, and 8 wt%). Composites were fabricated via the hand lay-up method with an optimal fiber loading of 15 wt%. The effects of chemical treatments were analysed through X-ray diffraction (XRD), Fourier-transform infrared spectroscopy (FTIR), and mechanical tests, including tensile, flexural, and impact testing. Results revealed that crystallinity index and crystalline size increased up to 6 wt% KOH treatment, attributed to the removal of surface impurities and partial elimination of hemicellulose and lignin, as confirmed by FTIR analysis. Correspondingly, mechanical strength improved up to 6 wt%, beyond which degradation occurred, likely due to polymer chain damage. Overall, composites reinforced with 6 wt% KOH-treated jute fibres exhibited the best mechanical performance, demonstrating the effectiveness of controlled alkali treatment for high-performance biocomposite development.

Keywords Biodegradable · Jute fibre · Potassium hydroxide (KOH) · X-ray diffraction (XRD) · Fourier-transform infrared spectroscopy (FTIR)

D. P. Sahu · S. C. Mohanty (✉)
Department of Mechanical Engineering, National Institute of Technology, Rourkela, Odisha, India
e-mail: scmohanty@nitrkl.ac.in

R. Das
Department of Physics and Astronomy, National Institute of Technology, Rourkela, Odisha, India

J. K. Prusty
Department of Mechanical Engineering, Madanapalle Institute of Technology & Science, Andhra Pradesh, India

H. A. C. Lopera et al. (eds.), *Natural Fibers and Biocomposites 2026*, The Minerals, Metals & Materials Series, https://doi.org/10.1007/978-3-032-13577-3_5

Introduction

The increasing use of dangerous chemicals and non-degradable compounds in our daily lives is a cause for concern, and it is imperative that we modify our lifestyle by replacing synthetic non-degradable substances with degradable ones [1, 2]. Driven by environmental concerns, researchers are working to develop materials that can replace synthetic materials with natural ones, thereby reducing the use of synthetic materials. Scientists are making advances in science and technology to produce engineering materials that are both environmentally friendly and beneficial to people [2–5]. Natural fiber-reinforced composites have attracted global interest due to their lighter weight, greater strength, and durability compared to their synthetic counterparts. They are also cost-effective, have a high strength-to-weight ratio, and require less energy during fabrication. In the 1960s, there was a growing demand for stiffer, stronger, and lighter materials in industries such as aerospace, construction, leisure, transportation, textiles, and sports. As a result, natural fiber-reinforced composites have become an increasingly popular choice for meeting these demands while also promoting sustainability [2–7]. There are two types of composites, biodegradable and non-biodegradable. In nowadays, lots of composites are prepared using synthetic fibers, which are non-degradable and have negative effects on our environment. But it was found that the composites using natural fibers are biodegradable, which motivates scientists to use natural fibers to make different composites. In general, fiber is a hair-like strand of material that is extremely long in comparison to its width. And it is made up of polymers, in which molecules are arranged in chains. These fibers are used in various fields for lots of activities, starting from human digestive to textile industries. On the basis of length, fibers are of two types, staple and filament fibers. The staple fibers have a short length, while filament fibers are long. But generally, fibers are classified as natural and artificial. In this project, natural fibers are used. Because these are easily available, biodegradable, having good strength and physical properties, the composite will be recyclable. Due to these properties, NFs are used in different forms like strands, textiles, rope etc. Also, due to this reason, it plays an important role in reinforcing agents for biocomposites and the applications of these biocomposites are found in various fields such as automobile, aerospace, home appliances, sports, building construction, etc. [8–11]. Natural fibers are three types: animal fiber, plant fiber, and mineral fiber. Further, plant fibers are classified into different types: leaf, bast, fruit, seed, wood, grass, and stalk. Here jute fiber is used as a reinforcement element for the fabrication of polymer composites, which comes under the bast fiber. Figure 1 illustrates the various classifications of fibers [8, 10, 12].

From the above study, it was noticed that jute fiber (Corchorus capsularis)-reinforced epoxy-based biocomposites were developed to enhance the mechanical strength of eco-friendly and biodegradable materials. However, the effect of alkali treatment on dewaxed jute fibers using different KOH concentrations (2wt%, 4wt%, 6wt%, and 8wt%) had not been extensively explored with respect to their crystalline and mechanical behavior. So, the author(s) of the present work bridge this gap and

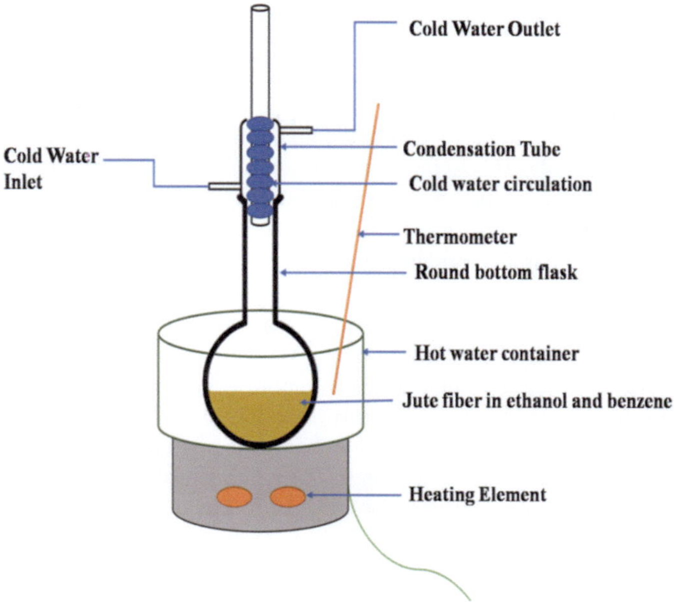

Fig. 1 Experimental setup for dewaxing

prepare jute/epoxy composites with 15wt% fiber loading using raw, dewaxed, and chemically treated fibers through the hand lay-up technique. The prepared composites were characterized by XRD, FTIR, and mechanical testing (tensile, flexural, and impact) to evaluate the influence of alkali treatment. It was found that crystallinity index, crystalline size, and mechanical strength improved up to 6wt% KOH treatment due to the removal of wax, pectin, hemicellulose, and lignin, whereas further increase in concentration led to a decrease, possibly due to polymer chain degradation.

Materials and Methods

The jute fiber used here is taken from the Karanjia market in Odisha. These collected raw fibers are of 150–360cm in length and have a diameter of 50–280μm approximately. These long jute fibers are cut in sizes of 1cm, which are used as raw material. From Himedia Laboratories Pvt. Ltd, Mumbai, the chemicals, Benzene (C_6H_6), Ethanol (C_2H_6O), Hardener (K-6), and Epoxy (L-12) are collected, which are required in this project.

Dewaxing Process

Dewaxing is a process that is used to purify and modify the surface of raw fibers. This is done by introducing a solvent made up of ethanol and benzene in a 1:2 ratio. The purpose of this solvent is to remove fatty substances like pectin and wax, as well as other impurities that are present on the surface layer of the fiber. These impurities can affect the adhesion between the fiber and matrix, which is why it is important to remove them. The apparatus used in this process includes several components. There is a heating element which is used to heat the container that holds the round bottom flask. There is also a temperature variator, which allows for precise control over the temperature inside the flask. A thermometer is used to measure the temperature inside the flask. Cold water circulation is used for vapour condensation. The round bottom flask contains the extractor solvent and raw fiber. It is placed in a container filled with normal water so that more than half of the beaker is covered by its outer side. This container is heated by the heating element. The heating process involves consecutive 2-h heating and 2-h cooling periods for up to 12 h. The temperature inside the flask is carefully maintained between 30°–65° C. After this, the fibers are extracted from the flask and washed in tap water to remove any remaining solvent or impurities. The fibers then go through a drying process in a vacuum oven at a temperature of 80° for 24 h. This ensures that all moisture is removed from the fibers. After this process, we get dewaxed jute fibers that are free from impurities and ready for use. The experimental setup for dewaxing is shown in Fig. 1.

KOH Treatment

After the dewaxing process, the cellulose of fibers is more active than before and it helps to crosslink with the matrix. Here for alkaline treatment, potassium hydroxide (KOH) solution is taken. To make treatment in different ratios, i.e., 2, 4, 6, and 8 wt%, the solution of KOH is mixed with DI (deionized) water respectively in separate beakers, where DI water is the purified water that has almost all its mineral ions removed. As an example, for 2 wt%, DI water is taken 196gm and KOH is taken 4gm. Similarly, for 4 wt%, DI water is taken 192gm and KOH is taken 8gm. After adding DI water and KOH, the solutions are mixed by the help of a magnetic stirrer for 2min. In these mixed solutions, 15gm of dewaxed jute fiber (DJF) is put for 1h in such a way that all fibers are in the solution. After 1h, all the fibers are extracted from beakers and washed thoroughly in water individually and then these rinsed fibers are placed in the vacuum oven in 80°C for 12h. After this process, finally, the 2, 4, 6, and 8 wt% KOH alkali-treated jute fibers are ready for sample preparation. The schematic diagram for composite fabrication is shown in Fig. 2.

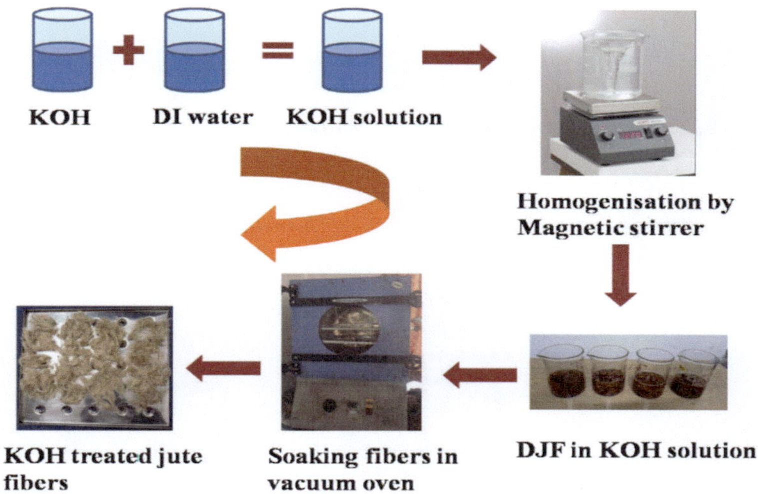

Fig. 2 KOH (potassium hydroxide) treatment of DJF

Preparation of Sample

By the hand lay-up method, samples of raw and dewaxed with physically treated samples are prepared. The process of preparing the composite in this method is one of the easy methods and it is cheap; due to these features, this method is used. To give a specific shape to the sample of composite, a (170 × 60 × 3) mm^3 dimension mold is prepared. Before fabrication of the sample, the complex bonds of the epoxy matrix is break by using the high-intensity ultrasonic processor for 1 h to convert in to a homogeneous matrix. Then this epoxy is made free from bubbles and put in the vacuum oven at 80°C for 6 h. After this, a matrix is prepared by adding this epoxy and hardener with a ratio of 10:1 and mixing continuously by the stirrer for 5 min, which is called resin. Here the hardener is mixed because it increases the rate of the curing process. The resin is further added with the jute fiber and mixed continuously for 5 min. by the stirrer to make blend for the fabrication of the sample. Then this blend is poured in to the mold. For easy removal of the composite slab from the mold, it is sprayed with a heavy-duty silicon spray before the blend is put into the mold. After putting into the mold, a plastic sheet is used to cover the blend in such a way that there should be no air bubbles left inside it. The removal of air bubble is made by rolling the iron roller homogeneously on them. After covering the plastic sheet, a glass pane is placed on it and a load of 40 to 60 kg is applied over it for 24 h to extract the extra resin and air bubbles left inside the blend. And after the removal of the composite slab, it is cut into the different required sizes for further experiments with the help of a zig-saw machine. The steps followed for composite fabrication are shown in Fig. 3.

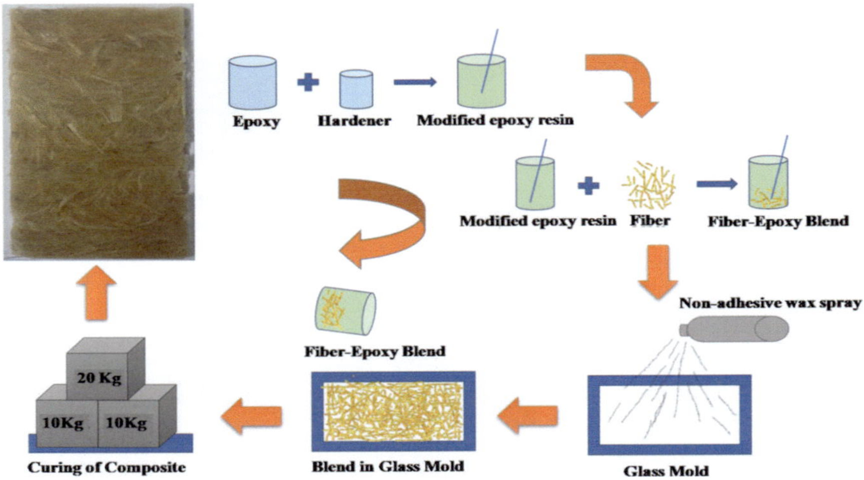

Fig. 3 Steps for composite fabrication

Characterizations

X-Ray Diffraction (XRD)

XRD is a non-destructive interference technique which gives information about the arrangement of atoms in the sample. To know the difference of atomic arrangement in row and dewaxed jute fiber, XRD patterns of both are used. These are obtained from the XRD using Cu Kα as the source of wavelength (λ) 1.54 Å at a scanning rate of 5°/min with applying 40 kV voltage and 40 mA current and a step size of 0.05°. The XRD patterns of the samples are obtained within 2θ range 5° to 40°. The image of X-ray spectrometer is shown in Fig. 4. For the calculation of crystallinity index of the fiber, Segal established a formula which is given by:

$$\text{Crl} = \left(\frac{i_{cr} - i_{am}}{i_{cr}}\right) \times 100\% \tag{1}$$

where $\boldsymbol{i}_{\text{cr}}$ is the intensity of cellulose peak, and $\boldsymbol{i}_{\text{am}}$ is the intensity of diffracted radiation from amorphous material.

In the sample, another important parameter is crystalline size, which is determined by the formula given by Scherrer:

$$\text{CrS} = \frac{R\lambda}{\mu \cos\theta'} \tag{2}$$

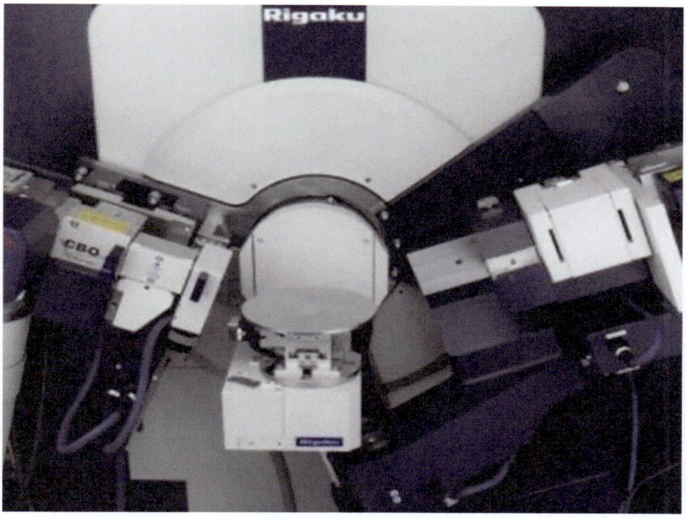

Fig. 4 Optical image of X-ray spectrometer

where R is Scherrer's constant equal to 0.94, θ is the peak position of intensity (maximum) found from the diffracted radiation, and μ stands for FWHM.

Fourier Transform Infrared Spectroscopy (FTIR)

The utilization of FTIR spectra is to identify the compounds present in the sample. In the FTIR spectrometer, samples are used in the pellet form, which is made by pressing the mixture of KBr binder and the powder of grinded fiber sample by a hydraulic presser. The mean percentage for transmittance of raw and dewaxed jute fiber is found between the wavelength scales of 4000–400 cm^{-1}, which lies under the near-infrared region. The mechanism of FTIR is shown in Fig. 6 and the image of FTIR spectrometer is in Fig. 5.

Flexural Strength

Flexural test is used to know the flexibility of the sample. It is also known as three-point bending test. Here the sample is cut into the required dimension and load is applied on its mid-point with a simple beam with a span-depth ratio of 16:1 until the sample's maximum capacity. The flexural strength is found out by the equation,

$$\sigma_{(\max)} = \frac{3F_{\max}L}{bh^2},$$

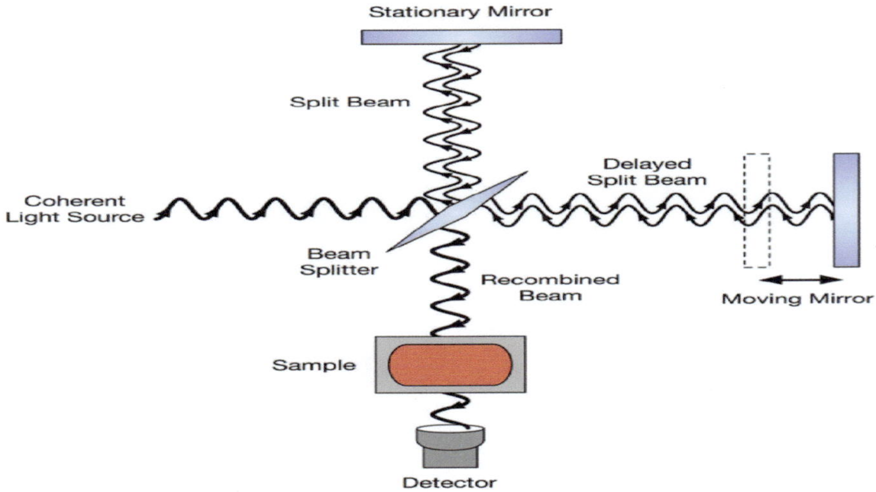

Fig. 5 Fourier transform infrared mechanism

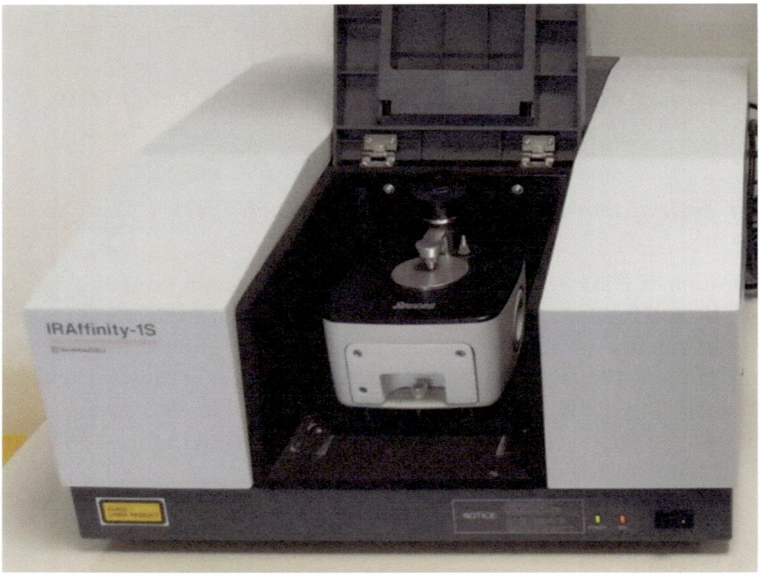

Fig. 6 Fourier transform infrared spectroscope

where F_{max} represents the maximum load (N) at the mid-point,
L is the gauge length (mm),
b is the width of the specimen (mm),
h is the thickness (mm) of the specimen.

Tensile Strength

This test is used to know the strength of the sample. In this test, the sample of the required dimension is placed between two fixtures called grips, which clamp the material and the weight is applied to the sample until it breaks down, gripped at one end while the other end is fixed. The optical image composite specimen in UTM for flexural test is shown in Fig. 7. The tensile strength is calculated using the formula,

$$\sigma_t = \frac{F_t}{b \times h},$$

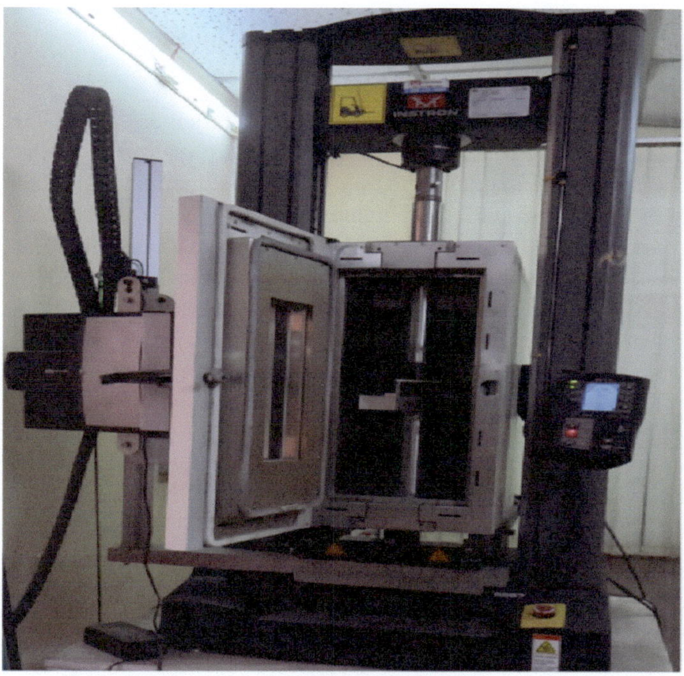

Universal flexural and tensile testing machine(UTM Instron-5967, UK)

Fig. 7 Composite specimen in UTM for flexural test

where F_t is the maximum load in MPa,
b is the width of the specimen (in mm),
h is the thickness of the specimen (in mm).

Impact Strength

Impact strength is the ability to endure a suddenly given fracture under stress. It can also be defined as the maximum capacity of a sample to oppose the suddenly applied force by any reason. And it is calculated in the form of energy. The impact strength of a material depends on its volume, force distribution, and modulus of elasticity. Similarly, in the case of biocomposites along with these factors, it is very much affected by the number of fibers present in that sample. Impact strength is usually quantified using the IZOD test or Charpy tests.

Results and Discussion

X-Ray Diffraction (XRD)

The information about crystallinity size and crystallinity index of the RJF, DJF, and other chemically treated fibers is found in the XRD plots in Fig. 9. As we know, the cellulose present in natural fibers is a semicrystalline material which means it contains both crystalline and amorphous regions. So in the XRD plots, there are mainly three peaks, middle one is the peak of cellulose with the miller indices [002] located at 22.25^0 and the other two peaks with miller indices [101] and [040] located at 15.72^0 and 34^0, respectively, are due to presence of cellulose. And the minimum intensity peak between two crystalline peaks present at 18.71^0 represents the non-cellulosic amorphous peak. The crystallinity index and crystalline size of all samples are given in Tables 1 and 2, respectively.

Due to the dewaxing process of RJF, the cementing layer present on the surface of jute fiber contains wax and other unwanted surface impurities is removed, which have the amorphous phase. Because of this removal of the amorphous phase, the crystallinity index of DJF increased from 63.79% to 68.91% and crystalline size is increased from 2.34nm to 2.62nm. The chemical treatment using KOH is performed by differing the wait % of KOH in DI water solution (i.e., 2wt%, 4wt%, 6wt%, and 8wt%). By analyzing the XRD patterns, it was found that the crystallinity index and crystalline size gradually increase up to 6 wt%, i.e., 75.35% and 3.05nm, respectively, and after this, it decreases. The reason is when we increase the wt% of KOH in solution up to 6 wt%, the polarized polar groups are oriented along the axis of crystalline cellulose and bond chain strength increases which increases the crystallinity index and size. But when we further increase the ratio of KOH content in solution,

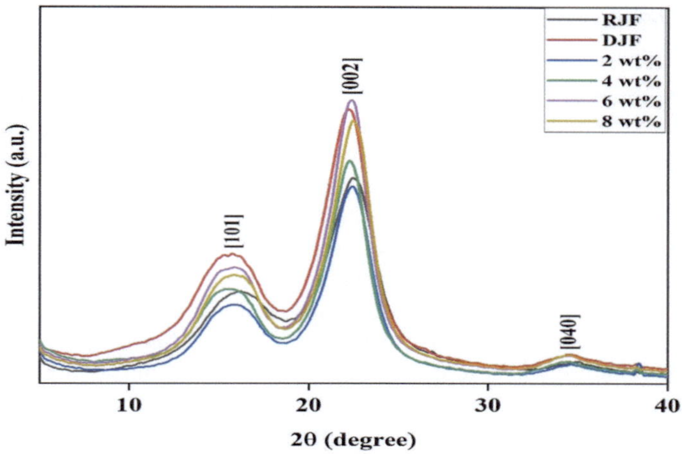

Fig. 8 XRD patterns of RJF, DJF, and chemically treated jute fibers

Table 1 Crystallinity index of RJF and treated jute fibers

Sample name	$I_{002} = I_{cr}$	I_{am}	CrI (%)
RJF	11,379.6572	4120.16271	63.79
DJF	14,849.9743	4615.727	68.91
2 wt%	10,973.5471	3021.6783	72.46
4 wt%	12,269.264	3260.4156	73.42
6 wt%	15,321.2455	3775.5244	75.35
8 wt%	14,285.8208	3727.7067	73.9

Table 2 Crystalline size of RJF and treated jute fibers

Sample name	β (in radian)	θ (in degree)	CrS (in nm)
RJF	3.51359	11.2425	2.34
DJF	3.13824	11.125	2.62
2 wt%	2.9343	11.216	2.8
4 wt%	2.75713	11.1385	2.98
6 wt%	2.69229	11.2035	3.05
8 wt%	2.81392	11.2295	2.92

the cellulose polymer chains start to degrade, due to which the crystallinity index and size decrease [3]. We can observe this in case of 8 wt%.

Fourier Transform Infrared Spectroscopy (FTIR)

The FTIR patterns of raw and treated jute fibers are shown in Fig. 10. The spectra is taken between the middle infrared region. In this spectra, the region is divided in two parts: one is diagnostic region and another one is fingerprint region. The diagnostic region is from wave number 4000 cm^{-1} to 1500 cm^{-1} and the fingerprint region is between wave number 1500 $^{cm-1}$ to 400 cm^{-1}. In natural fibers, the impurities like cellulose, hemicellulose, and lignin are mainly responsible for infrared patterns in FTIR. The FTIR spectra give information about the chemical groups present in a specimen. In Table 3, the functional groups are mentioned corresponding to the peaks with their wave numbers. The peak at 3577 cm^{-1} shows the $-OH$ stretching vibration of cellulose which represents the presence of cellulose in jute fiber. And the peak at 2921 cm^{-1} represents the $-CH_2$ and $-CH_3$ stretching of cellulose and hemicellulose. The sharp peaks at 1730 cm^{-1} is due to the $C = O$ stretching vibration of carboxylic acid or ester present in hemicelluloses, but in the KOH-treated jute fibers, this peak is absent because of the removal of hemicellulose due to alkali treatment. The removal of hemicellulose increases the attachment between fiber and matrix, which increases the mechanical properties of treated composites. Near this peak, water absorption peak is there at 1650 cm^{-1}. Along with this, there are three peaks associated with different components of lignin at 1597 cm^{-1}, 1507 cm^{-1}, and 1453 cm^{-1}. These peaks remain unchanged.

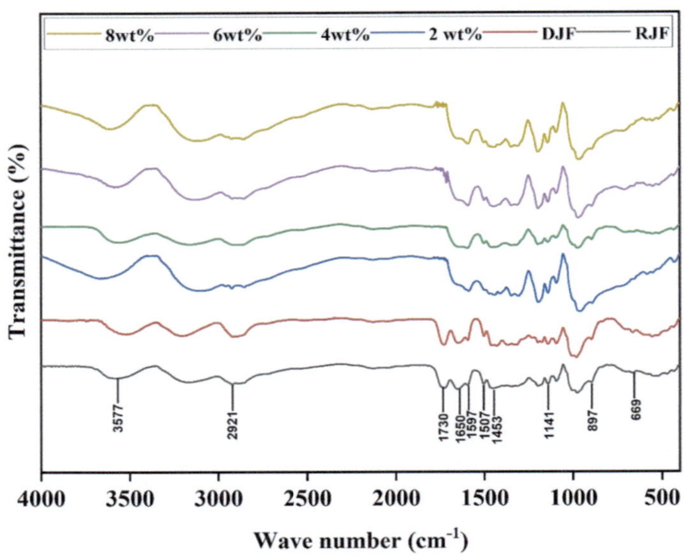

Fig. 9 FTIR spectra of treated and untreated jute fiber

Table 3 FTIR peaks and corresponding function groups

Wave number(cm^{-1})	Corresponding functional groups
3577	$-OH$ stretching of cellulose
2921	$-CH_2$ and $-CH_3$ stretching of cellulose and hemicellulose
1730	$C = O$ stretching of Ester
1650	$H - O - H$ bending of absorbed water
1597	$C = C$ bending with the aromatic ring of lignin
1507	Benzene ring stretching in lignin
1453	$-CH_2$ bending in lignin
1141	$C - O - C$ anti-symmetric bridge stretching in cellulose and hemicellulose
895	$-CH$ glycosidic bond stretching
669	$-OH$ bending

Mechanical Properties Analysis

Effect of Fiber Loading on Mechanical Properties of the Composites

Mechanical testing is performed on fiber-loaded composites to determine the optimal fiber loading for the composites. The results, as shown in the histogram in Fig. 11, indicate that the mechanical strength of the composites, including flexural, tensile, and impact strength, increases up to a fiber loading of 15 Wt.%. This increase in strength is due to an increase in interfacial stress transfer between the reinforcement phase and the matrix phase. However, as the fiber loading in the composites continues to increase beyond 10 Wt.%, the mechanical strength begins to decrease. This decrease in strength is due to a decrease in the wettability of the fiber [14], which results in a decrease in interfacial stress transfer between the fiber and matrix. Additionally, as fiber loading increases, the void content in the composite also increases, which further decrease mechanical strength. Therefore, it has been determined that 15 Wt% fiber-loaded composites are optimal for further application.

Effect of Dewaxing and Alkali Treatment on Mechanical Properties of Jute Composites

Based on the results from the previous section, it was determined that a fiber loading of 15 weight percent provided the highest mechanical strength. To further optimize the mechanical properties of the composites, dewaxed and alkali-treated jute fibers were used to fabricate composites with varying levels of alkali treatment. The flexural, tensile, and impact strength of these composites were then measured and are shown in Fig. 12.

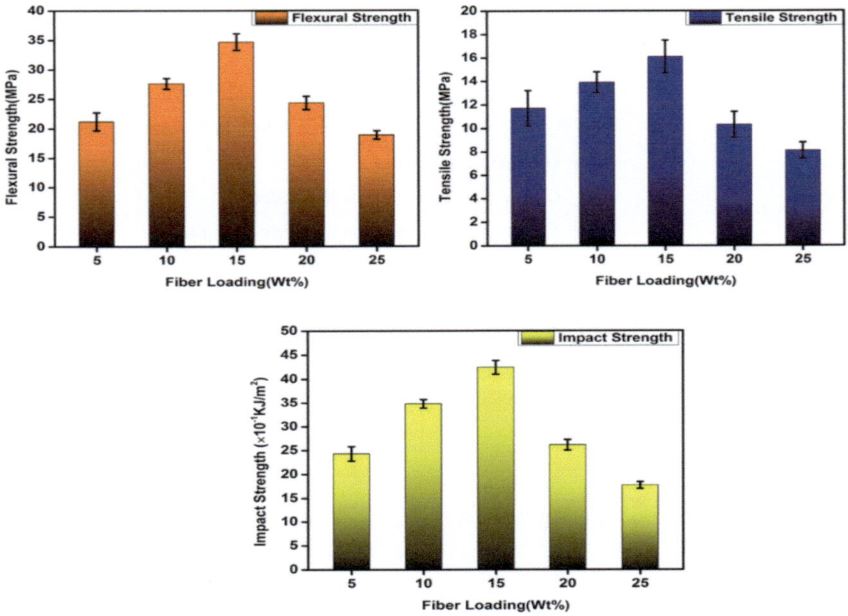

Fig. 10 Mechanical strength of RJF at different wt%

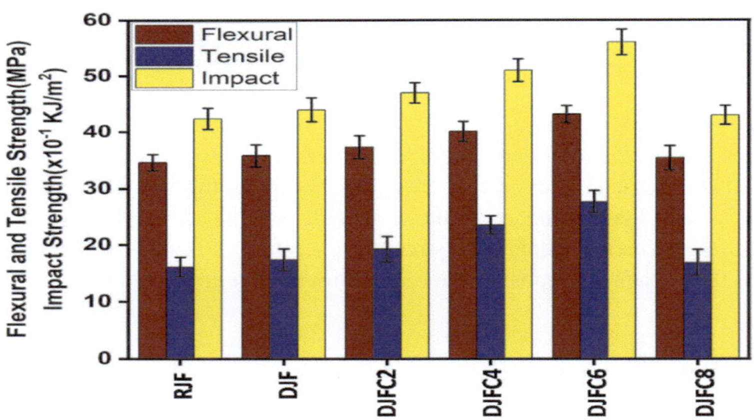

Fig. 11 Flexural, tensile, and impact strength of RJF, DJF, DJFC2, DJFC4, DJFC6, and DJFC8

After conducting mechanical testing on composites, it was discovered that the mechanical strength of DJF increased in several areas. The flexural strength increased by 9.41%, the tensile strength by 14.23%, and the impact strength by 6.52%. This increase in strength is due to increases in CrI and CrS, as well as the removal of impurities such as dust, wax, and other substances that can make the surface rough and hinder bonding with the matrix during composite fabrication. When DJF is treated with an alkali solution, the mechanical strength of the composites increases even further. This is due to the arrangement of polar bonds in a direction that increases CrI and CrS, as confirmed by XRD analysis. When DJF is treated with a 2 wt% KOH alkaline solution, the tensile strength of DJFC2 increased by 12.33%, the flexural strength by 36.52%, and the impact strength by 13.17%. When treated with a 4 wt% solution, the strength increases by 37.21, 78.46, and 29.94%, respectively; when treated with a 6 wt% solution, the strength increases by 51.06, 107.83, and 42.65%; and when treated with an 8 wt% solution, the strength increases by 11.76, 26.61, and 10.92% respectively. From the histogram, it is clear that DJFC6 shows the best mechanical performance due to effective stress transfer between fiber and epoxy resin and negligible void content in the composites.

Conclusion

In this project, we have synthesized the composites of jute fiber, reinforced with the epoxy matrix and along with the alkaline KOH treatment at different wt% to study the effect of this chemical treatment on composites' mechanical properties. From the works we have done above, it was found that the dewaxed jute fiber-reinforced epoxy composite with 6 wt% KOH treatment (DJFC6) has more crystallinity index, crystalline size, tensile strength, flexural strength, and impact strength than others. So 6 wt% KOH-treated DJF is more reliable for making composites. Surface morphology studies have shown that jute fibers treated with 6wt% KOH have increased adhesiveness, making them suitable for bonding. This treatment results in an optimal surface roughness that enhances the mechanical strength of composites prepared with these fibers. From XRD analysis, it was found that the crystallinity index and crystalline size increases up to 6 wt% KOH treatment and after that it decreases due to swelling of fiber. The FTIR plot shows that the content of hemicelluloses in fiber decreases by the KOH treatment. And from the mechanical testing, it was found that the mechanical strength is more in case of 6 wt% KOH-treated composite.

References

1. Ramesh M, Rajeshkumar L, Deepa C, Selvan MT, Kushvaha V (2021) Impact of Silane Treatment on Characterization of Ipomoea Staphylina Plant Fiber Reinforced Epoxy Composites Impact of Silane Treatment on Characterization of Ipomoea. J Nat Fibers. 00:1–12. https://doi.org/10.1080/15440478.2021.1902896
2. Shahidi S, Ghoranneviss M (2013) Sterilization of cotton fabrics using plasma treatment. In: Plasma Sci Technol. p 15. https://doi.org/10.1088/1009-0630/15/10/13
3. Dash C, Das R, Sahu DK, Upreti D, Patro TU, Bisoyi DK (2022) Investigation of dielectric and mechanical properties of pretreated natural sunn hemp fiber-reinforced composite in correlation

with macromolecular structure of the fiber. Biomacromol. https://doi.org/10.1021/acs.biomac.2c01370
4. Sun D, Stylios GK (2004) Effect of low temperature plasma treatment on the scouring and dyeing of natural fabrics. Text Res J 74:751–756. https://doi.org/10.1177/004051750407400901
5. Xu W, Mu B, Wang A (2017) Electrochimica Acta Three-dimensional hollow microtubular carbonized kapok fi ber/cobalt-nickel binary oxide composites for high-performance electrode materials of supercapacitors. Electrochim Acta 224:113–124. https://doi.org/10.1016/j.electacta.2016.12.032
6. Zille A, Oliveira R, Souto P (2014) Plasma treatment in textile industry. 1–34. https://doi.org/10.1002/ppap.201400052
7. Aminudin E, Khalid NHA, Azman NA, Bakri K, Din MFM, Zakaria R, Zainuddin NA (2017) Utilization of baggase waste based materials as improvement for thermal insulation of cement brick. MATEC Web Conf 103:1–8. https://doi.org/10.1051/matecconf/201710301019
8. Taj S, Munawar MA (2014) Natural fiber-reinforced polymer composites natural fiber-reinforced polymer composites
9. Asumani OML, Reid RG, Paskaramoorthy R (2012) The effects of alkali-silane treatment on the tensile and flexural properties of short fibre non-woven kenaf reinforced polypropylene composites. Compos Part A Appl Sci Manuf 43:1431–1440. https://doi.org/10.1016/j.compositesa.2012.04.007
10. Sgriccia N, Hawley MC, Misra M (2008) Characterization of natural fiber surfaces and natural fiber composites. Compos Part A Appl Sci Manuf 39:1632–1637. https://doi.org/10.1016/j.compositesa.2008.07.007
11. Rangappa SM, Parameswaranpillai J, Siengchin S (2022) Bioepoxy based hybrid composites from nano - fillers of chicken feather and lignocellulose Ceiba Pentandra, Sci Rep. 1–18. https://doi.org/10.1038/s41598-021-04386-2
12. Seki Y, Sarikanat M, Sever K, Erden S, Gulec HA (2010) Effect of the low and radio frequency oxygen plasma treatment of jute fiber on mechanical properties of jute fiber/polyester composite 11:1159–1164. https://doi.org/10.1007/s12221-010-1159-5
13. Al-maharma AY, Sendur P, Fiber SW, Polypropylene RH, Tusnim J, Jenifar NS, Hasan M (2020). Acoustic isolation of jute and kapok fiber reinforced polypropylene composites acoustic isolation of jute and kapok fiber reinforced polypropylene composites. https://doi.org/10.1088/1757-899X/932/1/012096
14. Das R, Dash C, Behera P, Bisoyi DK (2022). Influence of dewaxing on mechanical properties of kapok fiber- reinforced polymer composite influence of dewaxing on mechanical properties of kapok fiber-reinforced polymer composite. https://doi.org/10.1088/1755-1315/1086/1/012054

Effect of Natural Fiber-Reinforced Composite Material on Optimized Structures Under Impact Loads: A Computational and Experimental Approach

Edisson F. Ordonez, Henry A. Colorado, and Wilfredo Montealegre

Abstract Computational modeling of different materials, including natural fiber-reinforced composites, has developed in parallel with experimental research in this field. As the mechanical and physical properties of these materials become better understood, computer simulations have been developed that predict their behavior under different conditions and help engineers design more efficient and resilient composites. Furthermore, it is well known that fiber-reinforced composites are widely used in impact load dissipation. This, coupled with topology optimization methods, can drive new structural designs that improve the performance of energy dissipation systems. In this work, natural hemp fibers were used to manufacture a fiber-reinforced composite material as the first layer, and an ABS polymer core was used as a flexible energy dissipation material. Natural fiber-reinforced composites, combined with topology optimization of the ABS polymer structure, generated an efficient impact energy dissipation system, which was subsequently modeled using the finite element method (FEM). To validate the results, the computational models were compared with the experimental ones, using the mechanical properties of the hemp composite and the ABS.

Keywords Composite materials · Natural fibers · Finite element simulation · Structural impact · Topological optimization

---

E. F. Ordonez (✉) · W. Montealegre
Universidad Nacional de Colombia—Sede Medellín—Facultad de Minas, Medellín, Antioquia, Colombia
e-mail: eordonezo@unal.edu.co

E. F. Ordonez · H. A. Colorado
Universidad de Antioquia UdeA, Medellín, Colombia

H. A. C. Lopera et al. (eds.), *Natural Fibers and Biocomposites 2026*, The Minerals, Metals & Materials Series, https://doi.org/10.1007/978-3-032-13577-3_6

Introduction

The decarbonization of structural materials is pushing designers beyond traditional glass and carbon fiber composites toward bio-derived reinforcements with lower embodied energy and end-of-life burdens. Among natural fibers, hemp has emerged as a technically compelling option due to its favorable specific stiffness/strength, relatively stable supply chains, and compatibility with thermoset and thermoplastic matrices. Recent reviews summarize the mechanical performance of natural fiber composites (NFCs) across sectors and highlight their potential for load-bearing components when anisotropy and variability are properly managed [1–5].

For low-velocity impacts (up to 100 m/s) standardized methods, ASTM D3039/D3039M for tensile properties and ASTM D8101/D8101M for projectile penetration provide the experimental backbone to quantify properties and penetration resistance in a reproducible way [5, 6].

Computationally, finite element modeling (FEM) supports predictive assessment of stress/strain fields and failure initiation under dynamic loading. When coupled with topology optimization (TO), FEM enables material-layout designs that reduce mass and concentrate stiffness/energy-absorption capacity where it matters most. Foundational developments by Bendsøe & Sigmund and recent TO-for-AM methods demonstrate the viability of optimization-driven geometries for high-performance structures [7, 8].

Experimental

Materials and Laminate Fabrication

Unidirectional hemp fibers were infused with an epoxy system using load-assisted resin infusion on flat mold. The stacking sequence was to 0° [8] to emphasize longitudinal properties and simplify orthotropic identification. Post-cure: 80 °C for 2 h. Target fiber volume fraction (V_f): ~0.40 (validated by mass/density balance).

Tensile Testing (ASTM D3039/D3039M)

Rectangular specimens (190 × 20 mm) were manufactured using 50 mm gauges. With this test, maximum effort and longitudinal modulus of elasticity were extracted. Using the law of mixtures, orthotropic properties were obtained (E1, E2, G12, ν12, ν21).

Table 1 Experimental parameters

Sample	Description	Speed (m/s)
1	Structure 1	80
2	Structure 2	80
3	Equivalent (without OT) 1	80
4	Equivalent (without OT) 2	80
5	Reference (without hemp)	80
6	Structure 1	90
7	Structure 2	90
8	Equivalent (without OT) 1	90
9	Equivalent (without OT) 2	90
10	Reference (without hemp)	90
11	Structure 1	100
12	Structure 2	100
13	Equivalent (without OT) 1	100
14	Equivalent (without OT) 2	100
15	Reference (without hemp)	100

Impact Testing (ASTM D8101/D8101M)

Flat panels (90 × 90 mm) were clamped using a lateral fixation and impacted with cylindrical projectiles (≈10 g, Ø 7 mm) launched by a compressed air gun. Nominal velocities: 80, 90, 100 m/s are shown in Table 1 (actual measured with optical gates; high-speed imaging for diagnostics). Outcomes: penetration/no-penetration, deformation.

Computational Modeling

Orthotropic FE Model and Explicit Dynamics

Orthotropic properties (E1, E2, G12, ν12) calculated Section "Tensile Testing (ASTM D3039/D3039M)" are presented in Table 2. These properties are used to feed the material model used for finite elements in the COMSOL Multiphysics software. The integrated composite sheets are modeled with a polymer plate without optimization. In addition, the material that complements the energy dissipation system and on which optimization will be applied corresponds to ABS polymer. The impact is simplified using an equivalent point charge that is calculated considering the kinetic energy of the projectile. Figure 1 shows the applied load.

Table 2 Orthotropic properties of composite

Properties	
E1	1960
E2	771.4
E3	771.4
nu12	0.328
nu13	0.328
nu23	0.36
G12	284.8
G13	284.8
G23	183.8

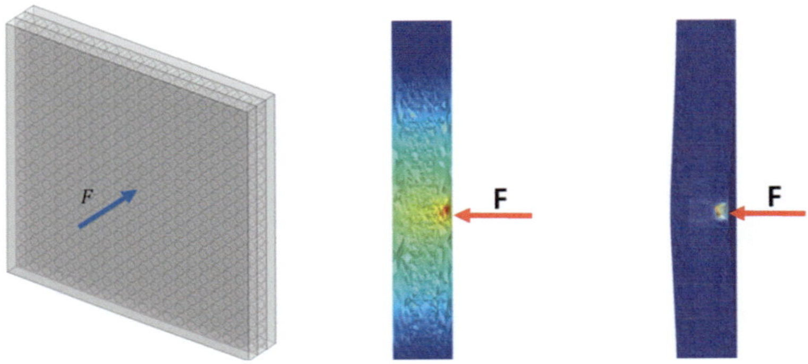

Fig. 1 Equivalent load applied

Topology Optimization

The topological optimization module of the COSMOL multiphysics software was implemented on ABS polymer plate, using static load equivalent to the impact, SIMP interpolation and Helmholtz type filter, with an Rmin defined by the mesh, Betha = 8 and Theta = 0.5. The results of the optimization are presented in Fig. 2.

Discussion

Optimization-driven Benefits

TO introduced material steering that (i) stiffens critical spans, (ii) offers alternate load paths, and (iii) provides sacrificial members for controlled damage, improving system performance. These trends mirror TO-for-AM studies in metals and polymers and

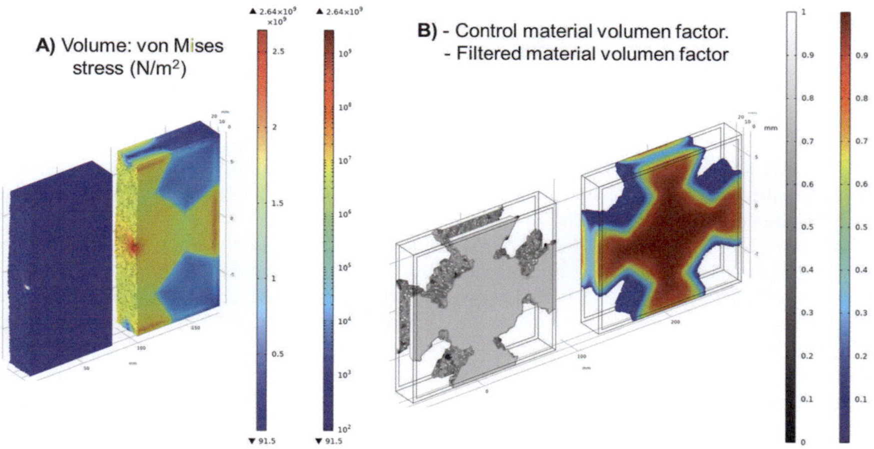

Fig. 2 Optimization results: **a** Von Mises stress, **b** control material volume factor

suggest clear potential for bio-composite panels manufactured with routing, lay-up tailoring, or hybrid layups [9].

Limitations and Future Work

(i) The failure criterion was calibrated with limited coupons; adopting progressive damage models with mixed-mode delamination laws would refine predictions. (ii) Environmental durability (humidity, UV) and interface treatments (alkali/compatibilizers) should be assessed for long-term stability. (iii) Manufacturability constraints for TO geometries in composites (minimum web/fillet sizes, drapability) merit deeper study.

Conclusions

- Topology-optimized panels at matched areal density reduced mass by 20–25% and delayed penetration by ~10%, improving stress redistribution and back-face response.
- The combined experimental–computational framework validates NFRPCs as candidates for lightweight, impact-resistant components when coupled with optimization-driven architectures.

References

1. Faruk O, Bledzki AK, Fink HP, Sain M (2012) Biocomposites reinforced with natural fibers: 2000–2010. Prog Polym Sci 37(11):1552–1596. https://doi.org/10.1016/j.progpolymsci.2012.04.003
2. Pickering KL, Efendy MGA, Le TM (2016) A review of recent developments in natural fibre composites and their mechanical performance. Compos Part A 83:98–112. https://doi.org/10.1016/j.compositesa.2015.08.038
3. Dittenber DB, GangaRao HVS (2012) Critical review of natural composites in infrastructure. Compos Part A 43(8):1419–1429. https://doi.org/10.1016/j.compositesa.2011.11.019
4. Tavares SS, de Mendonça Neuba L, Lopera HC, Monteiro SN, Meyers MA (2025) The embira bark fiber: a sustainable Amazon tape. Adv Compos Hybrid Mater 8(2):1–18. https://doi.org/10.1007/S42114-024-01170-4/TABLES/3
5. Safri SNA, Sultan MTH, Jawaid M, Jayakrishna K (2018) Impact behaviour of hybrid composites for structural applications: a review. Compos Part B 133:112–121. https://doi.org/10.1016/j.compositesb.2017.09.008
6. ASTM D3039/D3039M-17 (2017) standard test method for tensile properties of polymer matrix composite materials. ASTM International. https://doi.org/10.1520/D3039_D3039M-17
7. ASTM D8101/D8101M-24 (2024) Standard test method for measuring the penetration resistance of composite materials to impact by a blunt projectile. ASTM International. https://doi.org/10.1520/D8101_D8101M-24
8. Bendsøe MP, Sigmund O (2003) Topology optimization: theory, methods and applications. Springer. https://doi.org/10.1007/978-3-662-05086-6
9. Zhu J, Zhou H, Wang C, Zhou L, Yuan S, Zhang W (2021) A review of topology optimization for additive manufacturing. Chin J Aeronaut 34(1):91–110. https://doi.org/10.1016/j.cja.2020.09.020

Production of HA-PLA Filaments for Additive Manufacturing from Hydroxyapatite of Biogenic Origin

Sandra Milena Restrepo Arcila, Santiago Erazo, Marco Antonio Márquez Godoy, and Henry Alonso Colorado Lopera

Abstract The combination of calcium phosphates (CaPs), such as hydroxyapatite (HA), with polylactic acid (PLA) represents an effective strategy for the development of composite biomaterials with applications in bone tissue engineering. While CaPs provide bioactivity and osteoconductivity, PLA offers good mechanical properties and processability. This synergy facilitates the fabrication of customized scaffolds by additive manufacturing. In this context, HA was synthesized from eggshells and biogenic sulfuric acid, proposing a sustainable and low-cost way to obtain bioactive materials. The HA-PLA blends were processed by extrusion and characterized morphologically (using SEM) and mechanically. Strength, stiffness, and deformation properties were evaluated, and it was observed that the incorporation of HA significantly modified the mechanical behavior of PLA, suggesting an improvement in the bioactivity of the material without compromising its processability. These results demonstrate the potential of HA-PLA composite for applications in tissue engineering, especially in the fabrication of personal bone structures.

Keywords Additive manufacturing · PLA-HA filaments · Mechanical tests

Introduction

Additive manufacturing (AM) [1] is recognized as a versatile technology that enables the fabrication of parts with complex geometries without requiring molds or templates, while also offering the possibility of adapting designs to specific needs [2]. PLA has established itself as one of the most widely used filaments in 3D printing thanks to its low processing temperature, dimensional stability, adequate print quality,

S. M. R. Arcila (✉) · M. A. M. Godoy
Biomineralogy and Biohydrometallurgy Laboratory, Facultad de Minas, Departamento de Materiales y Nanotecnología, Universidad Nacional de Colombia–Medellín, Medellin, Colombia
e-mail: srestrepoar@unal.edu.co

S. M. R. Arcila · S. Erazo · H. A. C. Lopera (✉)
CCComposites Laboratory, Universidad de Antioquia UdeA, Medellín, Colombia
e-mail: henry.colorado@udea.edu.co

H. A. C. Lopera et al. (eds.), *Natural Fibers and Biocomposites 2026*, The Minerals, Metals & Materials Series, https://doi.org/10.1007/978-3-032-13577-3_7

and acceptable mechanical properties [3]. Among the advantages of AM is the possibility of using wastes as part of the raw materials, which can include hazardous materials [4], biochar [5], construction and demolition waste [6], tire rubber waste [7], and natural fibers [8]. Particularly, fused deposition modeling (FDM) [9] uses thermoplastic filaments, in which natural fibers have been used [8–10], reducing not only cost, but also making the solution more sustainable.

On the other hand, ceramic materials such as calcium phosphates (CaPs) are part of bone tissue and, due to their biocompatibility, have been synthesized and used in biomedical implants, with hydroxyapatite (HA) being the most common compound, especially as a coating on implantable devices [11]. However, its application in structural load environments is limited, which motivates its combination with polymer matrices such as PLA. The incorporation of HA into PLA allows the bioactivity and bone affinity of the ceramic to be exploited, while increasing the stiffness and compressive and flexural strength of the composite material. In PLA/HA mixtures with proportions between 0 and 15%, tensile tests have shown increases in the elastic modulus of approximately 2.3–3.5 GPa, as well as in the ultimate strength of 32.7 to 47.3 MPa with 15% HA [12]. However, at higher contents, porosity and the tendency to agglomerate reduce ductility and lead to more brittle fracture behavior.

This research focuses on the development of polylactic acid (PLA) filaments reinforced with hydroxyapatite (HA) in different concentrations. For their comprehensive characterization, scanning electron microscopy (SEM) was used to analyze the morphology and dispersion of the particles, in addition to mechanical tests to evaluate the behavior of the material under load conditions.

Materials and Methods

The materials, polylactic acid (PLA) pellets, are obtained from Estra SA in Medellín, and HA powder were obtained by chemical precipitation using a sulfuric acid leachate solution produced by a strain of *Acidithiobacillus thiooxidans* (*A.t*) (ATCC 13,977), which solubilizes sulfur (as a source of phosphorus), with calcium hydroxide obtained from eggshells. The precipitate was dried and characterized. To manufacture PLA-HA filaments, the granulated PLA was mixed manually for 5 min to achieve homogeneity, with different proportions of HA in varying percentages by weight (0%, 1%, 2.5%, 5%, and 10%). The homogenized mixture of each combination was fed into the extruder, where it was heat-fused, facilitating the efficient fusion of PLA and HA.

Results and Discussion

In the first stage of this research, the leaching of a low-grade phosphate rock was carried out using sulfuric acid of biogenic origin, produced by the *A.t* strain, with the purpose of obtaining mainly phosphoric acid [1, 2]. Subsequently, this acid reacted with calcium hydroxide, obtained from the calcination and subsequent hydration of eggshells [3–5], which allowed the synthesis by chemical precipitation of a calcium phosphate of the hydroxyapatite type [5, 6], according to the scheme shown in Fig. 1. The resulting material was dried until pH 7, and the powder obtained was incorporated into a PLA matrix in proportions of 1, 2.5, 5, and 10%, followed by an extrusion process at a temperature of 160 °C and a speed of 3 rpm and filaments were obtained from 130 mm approx. The morphological characterization of the filaments was performed by scanning electron microscopy (SEM), the results of which are presented in Fig. 2, while the evaluation of the mechanical properties is shown in Fig. 3.

Figure 2 shows SEM micrographs at 200X magnification of the fracture surfaces of the filaments. At lower HA contents (figure a), the particles are smaller and isolated, and the matrix exhibits a brittle fracture; the particles are dispersed more uniformly, with a relatively homogeneous distribution in the matrix. However, as the percentage of HA increases, a progressive increase and a marked tendency for the particles to agglomerate irregularly are observed, especially in Figure d, forming larger domains or clusters, suggesting a less homogeneous dispersion due to the high concentration.

Figure 3 shows the stress–strain curves for different percentages of HA. Using this data, we calculated the Young's modulus for each of the samples, obtaining: 2063.9 ± 4.5%, 983.72 ± 10%, 1201.7 ± 7.5%, 1110.9 ± 8%, and 1519.1 ± 10% MPa for 0, 1, 2.5, 5, and 10% HA, respectively.

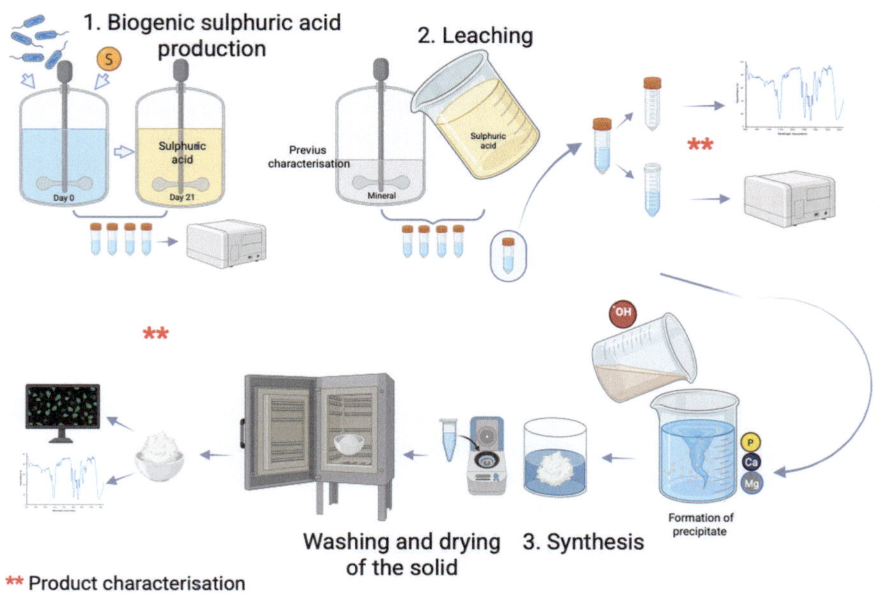

Fig. 1 Schematic figure of the process to obtaining HA

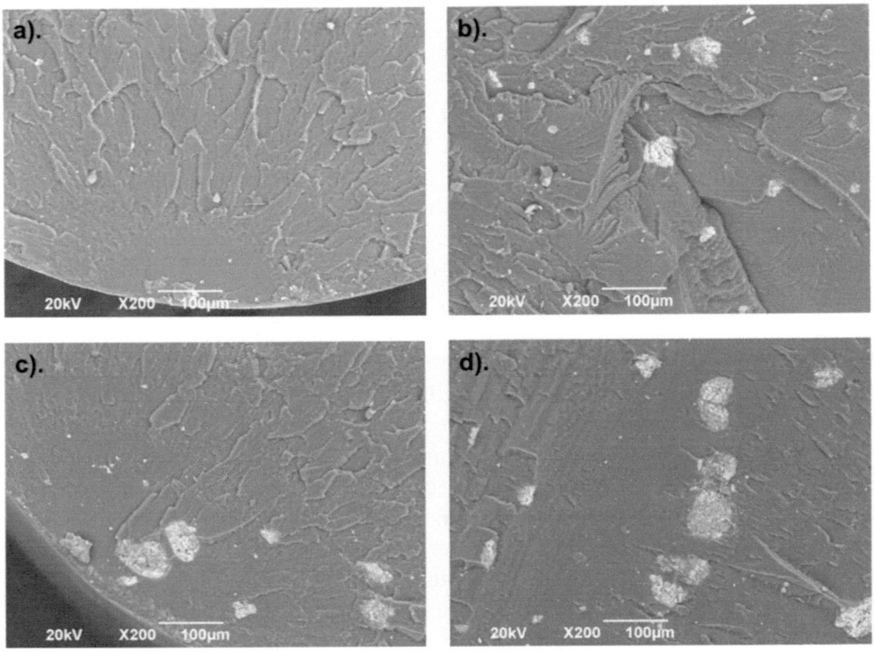

Fig. 2 SEM micrograph filaments in **a** 1% HA, **b** 2.5% HA, **c** 5% HA, and **d** 10%HA

The elastic modulus obtained for PLA has a relatively high initial stiffness due to its semi-crystalline nature. However, the addition of HA in low proportions (1–5%) produced a decrease in the elastic modulus; nevertheless, these show a high plastic deformation zone for 5, 2.5, and 1% HA, respectively. Meanwhile, for 10% HA, there is an increase in the elastic modulus without elastic deformation, suggesting that a direct fracture of the filament occurred.

Conclusions

Structurally, the presence of large HA agglomerates in the PLA matrix generates local heterogeneities, which can act as stress concentration points and cause a loss of ductility. At the same time, the higher fraction of ceramic reinforcement increases the overall stiffness of the composite, but with the risk of inducing more brittle behavior under mechanical stress, as evidenced by the reduction in strain at break in tensile studies.

The combination of PLA and HA is an effective strategy for developing composite materials with potential in biomedical applications, particularly in the manufacture of scaffolds and devices for bone tissue engineering. The resulting material combines the

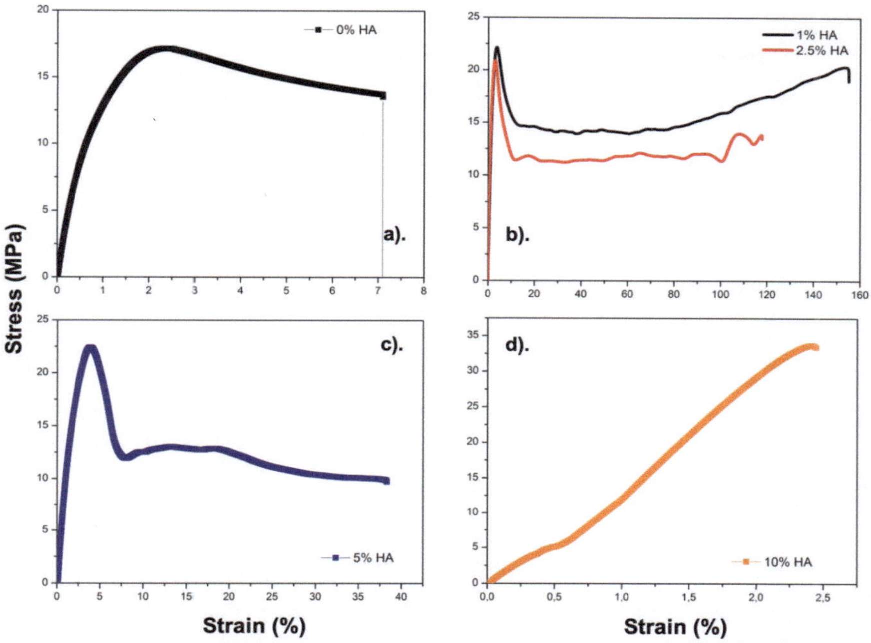

Fig. 3 Stress–strain curves for PLA filaments with HA contents of **a** 0%, **b** 1 and 2.5%, **c** 5%, and **d** 10%

bioactivity and cell affinity of HA with the processability and dimensional stability of PLA in 3D printing.

The incorporation of HA into the PLA matrix significantly modifies the mechanical behavior, reducing stiffness at low concentrations.

The microstructural evidence from SEM supports the values obtained from mechanical testing: as HA content increases, poor dispersion and particle agglomeration result in a loss of uniformity and greater mechanical variability. The combination of both observations indicates that there is an optimal range for the proportion of HA in PLA between 2.5–5%, in which a balance between reinforcement and structural homogeneity is achieved.

References

1. Kumar L, Pandey P, Wimpenny D (2019) 3D printing and additive manufacturing technologies. Singapore: Springer Singapore. https://doi.org/10.1007/978-981-13-0305-0
2. Colorado HA, Gutierrez-Velasquez EI, Gil LD, de Camargo IL (2024) Exploring the advantages and applications of nanocomposites produced via vat photopolymerization in additive manufacturing: a review. Springer Science and Business Media B.V. https://doi.org/10.1007/s42114-023-00808-z

3. Plamadiala I, Croitoru C, Pop MA, Roata IC (2025) Enhancing polylactic acid (PLA) performance: a review of additives in fused deposition modelling (FDM) filaments. Multidisciplinary Digital Publishing Institute (MDPI). https://doi.org/10.3390/polym17020191
4. Ordoñez E, Neves Monteiro S, Colorado HA (2022) Valorization of a hazardous waste with 3D-printing: combination of kaolin clay and electric arc furnace dust from the steel making industry. Mater Des 217:110617. https://doi.org/10.1016/j.matdes.2022.110617
5. Vergara LA, Perez JF, Colorado HA (2023) 3D printing of ordinary Portland cement with waste wood derived biochar obtained from gasification. Case Stud Constr Mater 18. https://doi.org/10.1016/j.cscm.2023.e02117
6. Colorado HA, Muñoz A, Neves Monteiro S (2022) Circular economy of construction and demolition waste: a case study of Colombia. Sustainability 14(12):7225. https://doi.org/10.3390/su14127225
7. Agudelo G, Cifuentes S, Colorado HA (2019) Ground tire rubber and bitumen with wax and its application in a real highway. J Clean Prod 228:1048–1061. https://doi.org/10.1016/j.jclepro.2019.04.353
8. Barreto G, Restrepo S, Vieira CM, Monteiro SN, Colorado HA (2024) Rice husk with PLA: 3D filament making and additive manufacturing of samples for potential structural applications. Polymers (Basel) 16(2):245. https://doi.org/10.3390/polym16020245
9. Paternina Reyes MJ, Unfried Silgado J, Santa Marín JF, Colorado Lopera HA, Espitia Sanjuán LA (2023) Cashew nutshells: a promising filler for 3D printing filaments. Polymers (Basel) 15(22):4347. https://doi.org/10.3390/polym15224347
10. Mwema FM, Akinlabi ET (2020) Basics of fused deposition modelling (FDM), pp 1–15. https://doi.org/10.1007/978-3-030-48259-6_1
11. Specification for Composition of Hydroxylapatite for Surgical Implants (2014) ASTM International, West Conshohocken, PA. https://doi.org/10.1520/F1185-03R14
12. Custodio CL, Broñola PJM, Cayabyab SR, Lagura VU, Celorico JR, Basilia BA (2021) Powder loading effects on the physicochemical and mechanical properties of 3D printed poly lactic acid/hydroxyapatite biocomposites. Int J Bioprint 7(1):112–122. https://doi.org/10.18063/ijb.v7i1.326
13. Calle-Castañeda SM, Márquez-Godoy MA, Hernández-Ortiz JP (2018) Phosphorus recovery from high concentrations of low-grade phosphate rocks using the biogenic acid produced by the acidophilic bacteria Acidithiobacillus thiooxidans. Miner Eng. https://doi.org/10.1016/j.mineng.2017.10.014
14. Calle-Castañeda SM, Márquez-Godoy MA, Hernández-Ortiz JP (2018) Solubilization of phosphorus from phosphate rocks with Acidithiobacillus thiooxidans following a growing-then-recovery process. World J Microbiol Biotechnol. https://doi.org/10.1007/s11274-017-2390-7
15. Khandelwal H, Prakash S (2016) Synthesis and characterization of hydroxyapatite powder by eggshell. J Miner Mater Charact Eng 04(02):119–126. https://doi.org/10.4236/jmmce.2016.42011
16. Suandork P, Hongsuwong M (2024) Synthesis of biocompatible hydroxyapatite from quail eggshell, oyster shell, and periwinkle snail shell. J Appl Res Sci Technol (JARST). https://doi.org/10.60101/jarst.2024.257693
17. Kareem Z, Eyiler E (2024) Synthesis of hydroxyapatite from eggshells via wet chemical precipitation: a review. RSC Adv 14. https://doi.org/10.1039/d4ra02198c
18. Wang LK, Vaccari DA, Li Y, Shammas NK (2005) physicochemical treatment processes: chemical precipitation. In: Wang LK, Hung Y-T, NK (eds) Handbook of environmental engineering. The Humana Press Inc., vol 3, pp 141–197

Eco UV Resin: Performance and Sustainability of a New Generation of Plant-Based Materials for 3D Printing

Victor Paes Dias Gonçalves, Ellen Bernardes de Lima, Verônica Scarpini Candido, Noam Tonini Simonassi, Carlos Maurício Fontes Vieira, and Felipe Lopes Perisse

Abstract Additive manufacturing has transformed the way parts are produced, enabling rapid prototyping, design customization, and efficient material usage. However, the widespread use of petroleum-derived resins raises environmental concerns due to their toxicity, volatile emissions, and improper disposal. In response, there is increasing demand for sustainable and biodegradable alternatives that maintain adequate technical performance. Plant-based resins have emerged as a promising solution, offering reduced environmental impact without compromising functionality. Eco UV Resin, formulated from soybean oil, represents an innovative green material designed for use with SLA/LCD 3D printing technologies. This study aimed to evaluate a UV-curable acrylate resin composed of bio-based components, focusing on its printability and post-processing behavior. Standardized samples were printed under fixed exposure and curing parameters. Mechanical, physical, and visual characterizations were performed. The results showed satisfactory performance for non-structural applications. Overall, Eco UV Resin appears to be a viable and environmentally friendly alternative for prototyping, design objects, and general-use prints, with future studies needed to assess long-term durability.

Keyword Photocurable resin · Sustainability · 3D printing

V. P. D. Gonçalves (✉) · E. B. de Lima · N. T. Simonassi · C. M. F. Vieira · F. L. Perisse
State University of the Northen Rio de Janeiro—UENF, Advanced Materials Laboratory, LAMAV, Campos Dos Goytacazes, Brazil
e-mail: victor_paesdias@yahoo.com.br

V. S. Candido
Materials Science and Engineering Program, Federal University of Pará, UFPA, Ananindeua, Brazil

H. A. C. Lopera et al. (eds.), *Natural Fibers and Biocomposites 2026*, The Minerals, Metals & Materials Series, https://doi.org/10.1007/978-3-032-13577-3_8

Introduction

Additive manufacturing (AM), and specifically VAT photopolymerization techniques such as SLA and LCD, have greatly advanced rapid prototyping and customized part production [1]. However, the widespread use of petroleum-derived UV-curable resins raises substantial concerns regarding toxicity, volatile organic compound (VOC) emissions, and end-of-life disposal issues [2]. Sustainable alternatives that preserve technical functionality while minimizing environmental harm are urgently needed [3].

Plant-based resins, particularly those derived from vegetable oils, offer a promising pathway toward greener VAT photopolymerization [4]. These bio-derived feedstocks are abundant, renewable, and modifiable to yield UV-curable monomers and polymers with properties comparable to conventional plastics [5]. Furthermore, adopting these bio-based resins aligns with broader sustainability goals, such as reducing carbon footprint, lowering toxicity, and promoting circularity in materials science [6, 7].

Recent studies have demonstrated the feasibility and potential of soybean oil-based resins in 3D printing applications [8, 9]. For example, formulations containing up to 80% bio-based acrylated or methacrylated epoxidized soybean oil have shown complete layer fusion and good mechanical tuning in SLA-printed [8, 9]. Novel one-step acrylation methods improve process efficiency and have enabled DLP printing of complex bio-based structures.

This work investigates Eco UV Resin, a plant-based, UV-curable acrylate resin formulated from soybean oil, intended for SLA/LCD 3D printing. We assess its printability, post-processing behavior, and performance across mechanical, physical, and visual domains, using standardized exposure and curing parameters.

Methodology

Material and Preparation

The Eco UV Resin is a UV-curable acrylate formulation primarily composed of polyurethane acrylate (30–60%) and acrylate monomers (10–40%). Bio-based fatty acid derivatives from soybean (epoxidized butyl esters, CAS 71302-80-2) contribute to its renewable content. The formulation also includes multifunctional diacrylates ($\approx$15%), which improve crosslinking density and mechanical strength. A photoinitiator system ($\approx$5%) ensures efficient UV curing, while stabilizing additives such as copper phthalocyanine pigments ($\approx$5%) provide long-term stability and color control. Overall, this balanced composition integrates renewable resources with reactive acrylates and curing agents to deliver high printability, durability, and performance.

Specimens were fabricated using an SLA/LCD system, the Anycubic Photon Mono 4 K (Fig. 1). Printing parameters were set to a layer height of 0.05 mm, with 6 base layers exposed for 50 s each, and subsequent layers exposed for 4 s. The building orientation was standard (edge-oriented). Post-processing involved two washing cycles in isopropyl alcohol (2 × 5 min), followed by air-drying for 15 min. Post-curing was conducted in a dedicated UV chamber using an array of light-emitting diodes (LEDs) with a peak emission wavelength of 405 nm. The specimens were positioned approximately 10 cm from the light source, and the measured average UV intensity at the sample surface was approximately 10 mW/cm^2. The total post-curing time was 20 min.

Fig. 1 SLA anycubic photon mono 4 K SLA/LCD

Confocal Laser Scanning Microscopy

Surface morphology and microstructural uniformity were analyzed using a confocal laser scanning microscope (Olympus LEXT OLS4000 3D). The system operates with a 405 nm laser and an optical setup enabling three-dimensional imaging with magnification. This method allowed quantitative topographic mapping, detection of defects such as voids and layer marks in the Eco UV Resin specimens.

Flexural Test

Flexural strength was determined to assess the mechanical integrity of the resin under bending stresses, relevant to functional applications. Testing followed ASTM D790 [10], with a velocity of 2 mm/min and 11 sample tested.

Compression Test

Compressive behavior was characterized according to ASTM D695 [11], using cylindrical specimens (cylinders 12.7 $\times$ 25 mm). The test was performed at a velocity of 1.3 mm/min until 10% strain or failure, with 11 sample evaluated. This method enables the determination of dimensional stability and load-bearing capacity under compressive stresses.

Density

Density was measured following ASTM D792 [12]. This parameter reflects both chemical composition and internal integrity, with deviations potentially indicating incomplete curing, void formation, or heterogeneity in the resin matrix. A total of three samples were tested.

Shore D Hardness

Surface hardness was evaluated using the Shore D scale according to ASTM D2240 [13]. Flat specimens with thickness $\geq$6 mm were tested, with measurements taken after 1 s of contact at $\geq$5 spaced points ($\geq$ 6 mm). Edges and regions with supported contact were excluded from analysis to ensure reproducibility. A total of seven samples were tested.

Izod Impact Test

Impact resistance was assessed using the Izod method in accordance with ASTM D256 [14]. Specimens contained a machined V-notch and were conditioned prior to testing with a pendulum calibrated to 22 J. Anomalous failures were discarded and replaced. A total of 11 samples were analyzed.

Accelerated Aging

Specimens were subjected to an accelerated aging protocol in accordance with ASTM G154 [15], using a UV/condensation chamber under controlled temperature and humidity conditions. The exposure lasted for 7 consecutive days, with alternating cycles of ultraviolet radiation and condensation to simulate environmental weathering. At the end of the process, the samples were removed, equilibrated at room conditions, and subsequently evaluated for changes in their physicochemical and mechanical properties.

Statistical Analysis

For the flexural and compression tests, data analysis was performed using the Student's *t*-test to assess statistical differences between groups. For the remaining quantitative tests (density, Shore D hardness, and Izod impact), the mean ($\bar{x}$) and the standard deviation (σ) were calculated to represent the central tendency and the variability of the material's properties.

Results and Discussion

The specimens were produced using an Anycubic Photon Mono 4 K printer, equipped with an LCD resolution of 4096 × 2560 pixels (4 K) and a horizontal resolution of 35 μm. Preliminary characterization by confocal laser scanning microscopy revealed an interlayer distance of 33.925 μm, confirming the dimensional precision and reliability of the resin specimens (Fig. 2).

The Eco UV resin specimens showed distinct behaviors under flexural and compressive loading (Table 2 and Fig. 3). In the as-printed condition, the average flexural strength was 64.7 ± 2.9 MPa, whereas the compressive strength reached 151.3 MPa. After UV aging, a significant reduction was observed in both properties ($p < 0.05$). Flexural strength decreased to 30.6 ± 4.5 MPa, representing a loss of

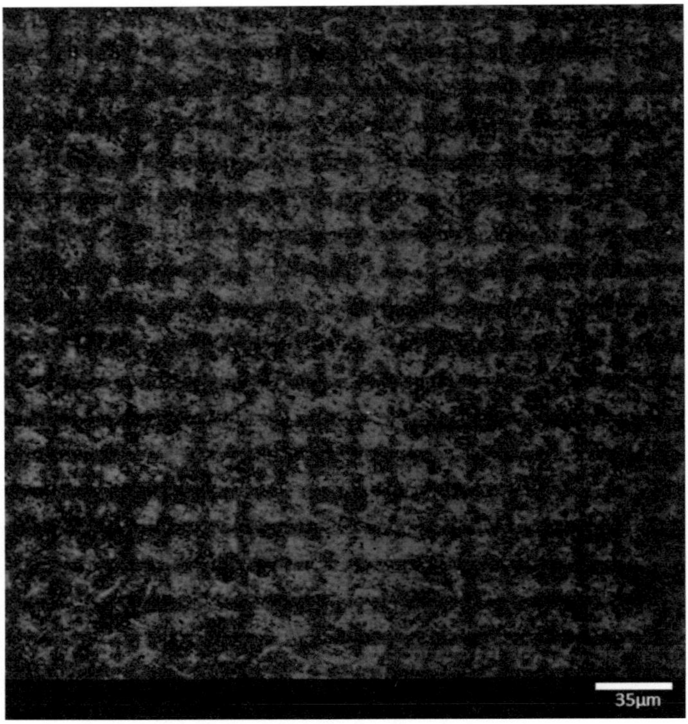

Fig. 2 Confocal laser scanning microscopy

more than 50%, while compressive strength dropped to an average of 104.8 MPa, corresponding to a reduction of approximately 30%.

The decline in flexural performance that UV exposure promoted degradation of the polymer network, possibly through chain scission and oxidative reactions, which compromised interlayer adhesion and reduced the material's ability to withstand tensile stresses generated during bending. In compression, the reduction was less severe, but still statistically significant, indicating that while the bulk structure retained some cohesion, microstructural embrittlement limited its resistance to high compressive loads.

These results contrast with the typical behavior of fully crosslinked photopolymers, in which UV post-curing usually enhances mechanical properties. In this case, the prolonged exposure appears to have induced photo-oxidative degradation rather than additional crosslinking, shifting the performance profile toward weaker and more brittle behavior. Such findings highlight the importance of optimizing curing and protective formulations when using bio-based resins to ensure mechanical stability under UV radiation.

The specimens exhibited an Izod impact resistance of 2.15 kJ/m^2 ($\pm$0.15), a value consistent with the range reported for photocurable SLA/LCD resins (Table 1)

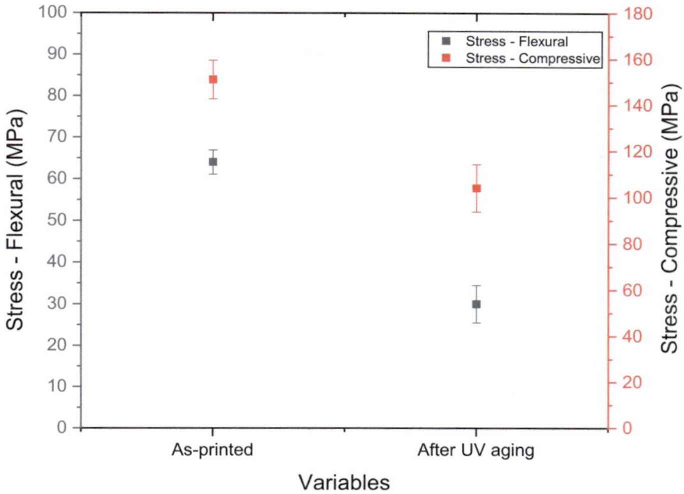

Fig. 3 Stress of flexural test and compressive test

[1, 16, 17]. The relatively low energy absorption during impact, combined with the sharp fracture profile, indicates a brittle failure mechanism characteristic of crosslinked acrylate-based polymers. This fracture behavior, governed by restricted chain mobility and limited plastic deformation, is typical of highly crosslinked networks. Compared with other resins in the literature (Table 2), the Eco UV resin demonstrated slightly higher impact resistance than dental resin (1.8 kJ/m^2) and ABS-like resin (1.7 kJ/m^2), but remained below more flexible standards, confirming its performance is closer to rigid photopolymers.

The Eco UV resin presented a density of approximately 1.15 g/cm^3, within the typical range for SLA/LCD resins, indicating that the partial substitution with bio-based feedstock did not compromise structural compactness [8]. The Shore D hardness of 79D confirmed a relatively rigid polymer matrix, comparable to engineering-grade photopolymers, providing adequate resistance to indentation and surface wear [8, 9].

The results confirm the suitability of Eco UV resin for additive manufacturing while highlighting its relevance as a bio-based alternative to conventional photopolymers. Derived from soybean oil, the resin reduces reliance on fossil resources, aligning with global initiatives for sustainable production and reduced environmental

Table 1 Flexural test and compressive test

	Stress-flexural	Stress-compressive	Test T
As-printed	64.7 ± 2.9	151.3	<0.05
After UV aging	30.6 ± 4.5	104.8	<0.05

Table 2 Results and comparison of literature

Materials	Impact IZOD resistence (kJ/m^2)	Ref.
Eco UV resin	2.15 ± 0.15	Reascher
Dental resin (3DCURE, São Paulo-SP, Brazil)	1.8 ± 0.6	[16]
ABS like (Elegoo, Shenzhen, China)	1.7 ± 1.1	[1]
Resin standart (3D LAB, Brazil)	1.3 ± 0.3	[17]
Splint cosmos (Yller, São Paulo-SP, Brazil)	1.4 ± 0.4	[17]

impact according the authors [12, 13]. By combining adequate mechanical performance with renewable feedstock, Eco UV resin represents a promising route for developing high-performance and eco-efficient materials within the framework of Industry 4.0.

Furthermore, the development of this bio-based resin aligns with the United Nations Sustainable Development Goals (SDGs), particularly Goal 9 (Industry, Innovation, and Infrastructure) and Goal 12 (Responsible Consumption and Production). Incorporating renewable raw materials into photopolymer resins demonstrates the potential of additive manufacturing to evolve toward circular economy strategies, integrating technological advancement with environmental responsibility.

Conclusion

This study demonstrated the feasibility of using Eco UV Resin, a plant-based acrylate resin derived from soybean oil, for SLA/LCD 3D printing.

By employing standardized printing and post-processing parameters, the findings confirm that the material can be successfully processed ensuring consistent specimen fabrication.

Overall, these results contribute to the understanding of bio-based resins as viable alternatives in additive manufacturing, while also providing a foundation for future studies aimed at optimizing their properties and expanding their applications.

Acknowledgements The authors thank the Brazilian agencies CNPq, process number 302976/2022-1, and FAPERJ, process number E-26/200.847/2021.

References

1. Dias Gonçalves VP, Vieira CMF, Simonassi NT, Perissé Duarte Lopes, F, Youssef G, Colorado HA (2024) Evaluation of mechanical properties of ABS-like resin for stereolithography versus aBS for fused deposition modeling in three-dimensional printing applications for odontology. Polymers 16:2921. https://doi.org/10.3390/polym16202921
2. Gil LD, de Camargo IL, Gutiérrez-Velásquez EI, Colorado HA (2023) Systematic overview of nanocomposites obtained by VAT photopolymerization techniques: a cost and life cycle assessment approach. UIS Ing 22:1–14
3. Pagac M, Hajnys J, Ma Q-P, Jancar L, Jansa J, Stefek P, Mesicek J (2021) A review of vat photopolymerization technology: materials, applications, challenges, and future trends of 3D printing. Polymers 13:598
4. Barkane A, Platnieks O, Jurinovs M, Kasetaite S, Ostrauskaite J, Gaidukovs S, Habibi Y (2021) UV-light curing of 3D printing inks from vegetable oils for stereolithography. Polymers 13:1195. https://doi.org/10.3390/polym13081195
5. Mendes-Felipe C, Isusi I, Gómez-Jiménez-Aberasturi O, Prieto-Fernandez S, Ruiz-Rubio L, Sangermano M, Vilas-Vilela JL (2023) One-step method for direct Acrylation of vegetable oils: a biobased material for 3D printing. Polymers 15:3136. https://doi.org/10.3390/polym15143136
6. Pakkanen J, Manfredi D, Minetola P, Iuliano L (2017) About the use of recycled or biodegradable filaments for sustainability of 3D printing: state of the art and research opportunities. Smart Innov Syst Technol 68:776–785
7. Hassan M, Mohanty AK, Misra M (2024) 3D printing in upcycling plastic and biomass waste to sustainable polymer blends and composites: a review. Mater Des 237:112558
8. Bodor M, Lasagabáster-Latorre A, Arias-Ferreiro G, Dopico-García MS, Abad M-J (2024) Improving the 3D printability and mechanical performance of biorenewable soybean oil-based photocurable resins. Polymers 16:977. https://doi.org/10.3390/polym16070977
9. Wang W, Hu Z, Lei W (1890) Research progress on vegetable oil-based UV-curing resins. Polymers 2025:17. https://doi.org/10.3390/polym17141890
10. ASTM D790-17 (2017) Standard test methods for flexural properties of unreinforced and reinforced plastics and electrical insulating materials. ASTM International, West Conshohocken, PA
11. ASTM D695-15 (2015) Standard test method for compressive properties of rigid plastics. ASTM International, West Conshohocken, PA
12. ASTM D792-20 (2020) Standard test methods for density and specific gravity (relative density) of plastics by displacement. ASTM International, West Conshohocken, PA
13. ASTM D2240-15 (2015) Standard test method for rubber property—durometer hardness. ASTM International, West Conshohocken, PA
14. ASTM D256-10 (2018) Standard test methods for determining the izod pendulum impact resistance of plastics. ASTM International, West Conshohocken, PA
15. ASTM ASTM G154-16 (2016) Standard practice for operating fluorescent ultraviolet (UV) lamp apparatus for exposure of nonmetallic materials. West Conshohocken (PA): ASTM International
16. Gonçalves VPD, Velasco DCR, Colorado HA, Vieira CMF, Perisse FL (2025) 3D printing in additive manufacturing in stereolithography: coconut shell powder additive. In: Lopera HAC et al. Natural fibers and biocomposites. The minerals, metals & materials series. Springer, Cham, pp 123–145. https://doi.org/10.1007/978-3-031-81190-6_9
17. Gonçalves VPD, Velasco DCR, Lucas J, Vieira CMF, Simonassi NT, Lopes FPD, Lopera HAC (2024) Avaliação do método de preparo do entalhe no ensaio de impacto de Izod para materiais de impressão 3D. In: 77° Congresso Anual da ABM; 2024; São Paulo. São Paulo: ABM, pp 1–7

18. Hafkamp T, van Baars G, de Jager B, Etman P (2018) A feasibility study on process monitoring and control in vat photopolymerization of ceramics. Mechatronics 56:220–241
19. Ryu K, Kim J, Choi J, Kim U (2022) The 3D printing behavior of photocurable ceramic/polymer composite slurries prepared with different particle sizes. Nanomaterials 12:2631

Characterization and FEM Orthotropic Modeling of Hemp Fiber-Reinforced Epoxy Matrix Composite Material

Edisson F. Ordonez, Wilfredo Montealegre, and Henry A. Colorado

Abstract This study explored the finite element modeling of a composite material made from epoxy resin and long hemp fibers arranged in a unidirectional orientation. Two methods were analyzed to estimate their mechanical properties: the law of mixtures method and the Mori–Tanaka method. These methods were used to calculate parameters such as elastic modulus, and shear strength, Poisson ratio, longitudinally and transversely. The result was a stiffness matrix that can be implemented in finite element software, in this case COMSOL, to simulate the composite under equivalent impact loads. This methodology not only demonstrated the viability of hemp as a natural reinforcement but also paves the way for research into other plant fibers in future developments. This work provides a method for future work with other types of natural fibers, which, due to their variable properties, make modeling with computational methods quite difficult.

Keywords Composites · Characterization · FEM simulation · Natural fiber composite

Introduction

The design and development of polymer matrix composites reinforced with natural fibers has received increasing attention in recent decades due to the need for sustainable, low-cost solutions with competitive mechanical performance compared to conventional composites based on synthetic fibers [1]. Natural fibers such as hemp, flax, sisal, or jute have shown significant potential in structural and semi-structural applications, mainly because of their low density, renewable availability, and favorable specific strength and stiffness [2, 3] particular, the use of long unidirectional

E. F. Ordonez (✉) · W. Montealegre
Universidad Nacional de Colombia—Sede Medellín—Facultad de Minas, Medellín, Antioquia, Colombia
e-mail: eordonezo@unal.edu.co

E. F. Ordonez · H. A. Colorado
Universidad de Antioquia—Facultad de Ingeniería, Medellín, Antioquia, Colombia

H. A. C. Lopera et al. (eds.), *Natural Fibers and Biocomposites 2026*, The Minerals, Metals & Materials Series, https://doi.org/10.1007/978-3-032-13577-3_9

fibers embedded in epoxy matrices enables laminates with pronounced orthotropic behavior, suitable for applications in sectors such as transportation, construction, and even lightweight ballistic protection.

To predict the mechanical behavior of such composites, it is necessary to employ micromechanical models that can estimate effective properties from the individual properties of the constituents and their volume fractions. Among the most widely used models are the rule of mixtures and the Mori–Tanaka method. The former is a classical and straightforward approach that assumes uniform stress or strain distribution along the reinforcement direction, providing acceptable estimations for the longitudinal Young's modulus [4, 5]. However, its accuracy decreases when predicting transverse and shear properties. In contrast, the Mori–Tanaka method, based on Eshelby's ellipsoidal inclusion theory, incorporates the fiber–matrix interaction and has proven to be more robust in estimating orthotropic properties, particularly in unidirectional fiber-reinforced materials with high fiber volume fractions.

The aim of this study is to compare the implementation of the rule of mixtures and the Mori–Tanaka model to determine the orthotropic properties of an epoxy composite reinforced with unidirectional natural fibers, using long hemp fibers as a case study [4, 5]. For this purpose, the theoretical foundations of both models, the governing equations, and the required input parameters are presented, along with the results obtained for the estimation of elastic moduli, Poisson's ratios, and shear moduli. These results are contrasted with reference experimental values obtained through tensile tests following ASTM D3039 [6]. The comparison allows assessing the suitability of each model in the characterization of natural fiber-reinforced composites and provides guidelines for their integration into finite element models aimed at structural analysis and advanced design applications [7].

Experimental

Materials and Manufacturing

The composite laminates were fabricated using long unidirectional hemp fibers embedded in an epoxy resin system. Fiber volume fraction was controlled to approximately 40%, ensuring uniform fiber alignment. Specimens were prepared and cut in accordance with ASTM standards for tensile mechanical testing [6].

Tensile Testing

Uniaxial tensile tests were carried out according to ASTM D3039/D3039M [6]. Five specimens were conditioned and mounted in mechanical grips with bonded tabs to minimize stress concentrations. Tests were performed under displacement control at a rate of 1 mm/min. Strain was measured using an extensometer over a gauge length of 50 mm. The parameters extracted included:

Estimation of Orthotropic Properties

To estimate the orthotropic elastic properties of the composite material with epoxy resin matrix reinforced with natural hemp fibers in unidirectional arrangement, two complementary micromechanical approaches were implemented: the law of mixtures (Voigt–Reuss) and the Mori–Tanaka model.

The simplified law of mixtures is based on the principle that the properties of the compound are a weighted combination of the properties of the matrix and the reinforcement, according to their respective volumetric fractions. The law of mixtures was applied in its classical form, using combinations of the Voigt (series) and Reuss (parallel) models to estimate the moduli in longitudinal and transverse directions.

The Mori–Tanaka model allows us to consider the mechanical interaction between the matrix and the fibers by mean of stress concentration factors. In this case, perfectly aligned cylindrical fibers were assumed. This method makes it possible to predict the orthotropic behavior of the material from the individual properties of the matrix and reinforcement, as well as the volumetric fraction of fiber [8, 9]. The model assumes that the fibers are perfectly aligned in a single direction and that the composite behaves in a linear elastic manner. Under these conditions, the compound exhibits orthotropic behavior, with different mechanical properties in the longitudinal direction (parallel to the fibers) and in the transverse directions. In general, it is obtained that E1, in the direction of the fibers, results from Voigt's law of mixtures (this is recovered from the Mori–Tanaka model for perfectly aligned fiber [4, 5].

Shear modulus:

$$G = \frac{E}{2(1+\nu)}.$$

Rule of Mixtures (RoM):

- Longitudinal direction (parallel to fibers):

$$E_1 = V_f E_f + V_m E_m$$

$$\nu_{12} = V_f \nu_f + V_m \nu_m$$

- Transverse and in-plane shear (inverse mixtures):

$$\frac{1}{E_2} = \frac{Vf}{E_f} + \frac{Vm}{E_m}$$

$$\frac{1}{G_{12}} = \frac{Vf}{G_f} + \frac{Vm}{G_m}.$$

If no better data is available: $\nu_{23} \approx \nu_m$
Mori–Tanaka (MT) Mean-Field Homogenization:
Effective constitutive tensor:

$$C_{eff} = C_m + V_f(C_f - C_m)A_{MT}$$

With A_{MT} derived from the tensor of Eshelby.
Coefficients of the cylindrical unidirectional case:

$$E_2 = E_m \frac{1 + AV_f}{1 - BV_f}$$

$$G_{12} = G_m \frac{1 + CV_f}{1 - DV_f}$$

$$\nu_{23} \approx \nu_m + EV_f.$$

$$Forunidirectional : E_1 \approx V_f E_f + V_m E_m.$$

The coefficients A, B, C, D, and E depend on the Eshelby tensor and the Poisson relations of the matrix.

The entire process for both the grade of mixtures and Mori–Tanaka was programmed and automated in MATLAB, allowing a direct comparison between the two methods under the same input parameters.

Results

The following parameters were considered as experimental parameters used as input for the MATLAB code: $V_f = 0.4$, modulus of elasticity of the matrix of $E_m = 2500$ MPa, modulus of elasticity of the compound of $E_{comp} = 1800$ MPa, and Poisson coefficients of $\nu_f = 0.25$, $\nu_m = 0.35$. The fiber modulus calculated was $E_f \approx 3187.5$ MPa.

From the experimental data of the individual phases (matrix and fibers), as well as the experimental properties of the tension test, the effective elastic properties of the compound were calculated using the mixing law methods and the Mori–Tanaka model, obtaining the results presented in Table 1.

Finite element simulations reproduced the tensile response with less than 5% deviation in stress–strain slope compared to experimental results.

Table 1 Comparison of effective properties. Law of mixtures and Mori–Tanaka

Property/method	Mixtures law	Homogenization
E1	1960.00	1960.00
E2	1958.80	1917.50
E3	1958.80	1917.50
nu12	0.31	0.31
nu13	0.31	0.31
nu23	0.35	0.35
G12	748.30	755.90
G13	748.30	755.90
G23	740.70	740.70

The tensile properties obtained in this study are consistent with those reported in the literature on hemp fiber-reinforced epoxy composites. The longitudinal modulus of 1800 MPa is comparable with the values reported by Gargol et al. [10], who observed a range of 1600–2000 MPa for chemically treated hemp fibers in epoxy matrices, and Sun et al. [11], who measured an average of 1780 MPa for physically treated hemp fibers.

Conclusions

- These results confirm that the behavior of the compound is highly anisotropic, with significantly higher stiffness in the longitudinal direction (parallel to the reinforcement).
- The results show an expected agreement in the longitudinal property (E1), since both methods coincide as they are perfectly aligned fibers. However, notable differences are observed in the transverse and shear properties, where the Mori–Tanaka model, when considering the redistribution of stresses between phases, offers more conservative and physically representative estimates than the law of mixtures.
- The comparison with the experimental data obtained from tensile tests showed a good correlation, which validates the use of the Mori–Tanaka model as a reliable predictive tool for materials with unidirectional fibers.
- Compared to the Mori–Tanaka model, the law of mixtures provides a simpler, although less precise approximation in situations where there is a marked difference between the properties of the phases or when the interaction between fibers cannot be neglected. Even so, in cases where the fibers are unidirectionally aligned and the volumetric fraction is moderate, the model provides results reasonably close to the values observed experimentally.

- This analysis allows for an initial orthotropic elastic characterization for the compound, suitable for use in numerical simulations with finite element software or as a basis for subsequent calibrations with experimental data.

References

1. Huzaifa M et al (2024) Exploring mechanical properties of eco-friendly hybrid epoxy composites reinforced with sisal, hemp, and glass fibers. J Mater Res Technol 33:2785–2793. https://doi.org/10.1016/J.JMRT.2024.08.189
2. Placet V (2009) Characterization of the thermo-mechanical behaviour of Hemp fibres intended for the manufacturing of high performance composites. Compos Part A Appl Sci Manuf 40(8):1111–1118. https://doi.org/10.1016/j.compositesa.2009.04.031
3. Mohanty Amar LK, Manjusri M, Drzal T (2005) Natural fibers, biopolymers, and biocomposites. CRC Press, Taylor & Francis Group
4. Camanho PP, Dávila CG, Pinho ST, Iannucci L, Robinson P (2006) Prediction of in situ strengths and matrix cracking in composites under transverse tension and in-plane shear. Compos Part A Appl Sci Manuf 37(2):165–176. https://doi.org/10.1016/J.COMPOSITESA.2005.04.023
5. Benveniste Y (1987) A new approach to the application of Mori-Tanaka's theory in composite materials. Mech Mater 6(2):147–157. https://doi.org/10.1016/0167-6636(87)90005-6
6. ASTM D3039/D3039M-17 (2017) Test method for tensile properties of polymer matrix composite materials. https://doi.org/10.1520/D3039_D3039M-17
7. Mulenga TK, Rangappa SM, Siengchin S (2025) Impact behavior of natural fiber composites: a comprehensive review on theoretical and computational modeling. Next Mater 8:100849. https://doi.org/10.1016/J.NXMATE.2025.100849
8. Ogierman W, Kokot G (2016) A study on fiber orientation influence on the mechanical response of a short fiber composite structure. Acta Mech 227(1):173–183. https://doi.org/10.1007/S00707-015-1417-0/METRICS
9. Kamarian S, Salim M, Dimitri R, Tornabene F (2016) Free vibration analysis of conical shells reinforced with agglomerated carbon nanotubes. Int J Mech Sci 108–109:157–165. https://doi.org/10.1016/J.IJMECSCI.2016.02.006
10. Gargol M, Klepka T, Klapiszewski Ł, Podkościelna B (2021) Synthesis and thermo-mechanical study of epoxy resin-based composites with waste fibers of hemp as an eco-friendly filler. Polymers (Basel) 13(4):503. https://doi.org/10.3390/POLYM13040503
11. Sun S, Yan Y, Zhou C, Pillay S, Ning H (2024) Mechanical performance of epoxy matrix composites reinforced by physically and chemically treated hemp fiber. Polym Compos 45(16):15113–15124. https://doi.org/10.1002/PC.28824

Thermal Characterization of Textile Waste and Its Feasibility for Application in Orthopedic Prostheses

A. F. Campanhão, V. P. D. Gonçalves, M. P. Oliveira, N. T. Simonassi, C. M. F. Vieira, and F. P. D. Lopes

Abstract Polymer composites reinforced with textile waste represent a sustainable and functional alternative to the high level of waste from the textile industry, one of the most polluting in the world. The incorporation of these materials into polymeric matrices has shown potential for applications in orthopaedic medicine, combining lightness, resistance and versatility. This research proposes the reuse of different types of linen, jeans, poplin, and helanca textile waste, assessing their viability for application in orthopaedic prostheses. Thermal and physical analyses were carried out using thermogravimetry (TGA), differential scanning calorimetry (DSC), and density by pycnometry, respectively. It was concluded that this study made it possible to observe the thermal performance of different fabrics and their potential application as reinforcements in composites and possible applications.

Keywords Polymer composites · Textile waste · Thermal analysis

A. F. Campanhão (✉) · V. P. D. Gonçalves · M. P. Oliveira · N. T. Simonassi · C. M. F. Vieira · F. P. D. Lopes
UENF—State University of the Northern Rio de Janeiro, LAMAV—Advanced Materials Laboratory, Campos Dos Goytacazes, Rio de Janeiro, Brazil
e-mail: angelicampanhao@gmail.com

M. P. Oliveira
e-mail: michel.oliveira@ufes.br

N. T. Simonassi
e-mail: noansimonassi@uenf.br

C. M. F. Vieira
e-mail: vieira@uenf.br

F. P. D. Lopes
e-mail: perisse@uenf.br

H. A. C. Lopera et al. (eds.), *Natural Fibers and Biocomposites 2026*, The Minerals, Metals & Materials Series, https://doi.org/10.1007/978-3-032-13577-3_10

Introduction

The increase in textile production and growth in clothing consumption have raised significant environmental concerns, particularly with regard to the generation of solid waste and increased pollution [1]. Recent research highlights the negative effects of the textile industry on aquatic and atmospheric ecosystems, associated with the excessive use of harmful chemicals during the various stages of the product life cycle [2].

The textile industry ranks among the five largest greenhouse gas emitting sectors globally, alongside energy, agriculture, land use and forestry, and industrial processes. It is estimated that textile processing is responsible for more than 1.2 billion tons of carbon emissions annually [3]. Global textile waste production is estimated at approximately 92 million tons per year, with forecasts pointing to an increase to 134 million tons by 2030 [4]. Currently, about 75% of this waste is sent to landfills, while only 25% undergoes reuse or recycling processes, with less than 1% actually being reintroduced into the production of new garments [5]. This scenario highlights the urgent need to develop and implement more efficient recycling technologies capable of mitigating the growing volume of waste and promoting the sustainable reuse of textile materials [6].

In this context, the proposal to valorize this waste through the development of polymer composites reinforced with textile materials emerges as a promising alternative. This approach combines the functional properties of polymers with the structural characteristics of textiles, creating an innovative and expanding field [7]. The integration of these materials results in multifunctional composites with high mechanical performance and broad adaptability to different applications. Their effectiveness has been demonstrated in several studies, with a special emphasis on their application in prostheses, providing efficient solutions for amputees or people with physical limitations [8, 9].

Currently, some authors have been investigating the thermal properties of textile fibers, using different approaches. Isola et al. [10] focused on the thermal analysis and pyrolysis decomposition mechanisms of chemically modified cotton fibers and concluded that silanization and sulfation-phosphorylation altered the thermal behavior of cotton, increasing stability and fire retardancy. These modifications expand its use in industrial thermal applications, while Kumar et al. [11] developed an innovative composite from textile waste, targeting specific mechanical, acoustic, and thermal properties. The authors concluded that natural biocomposites have good thermomechanical properties and controlled degradation, ensuring effective structural support. This reinforces their potential for comfortable and functional prostheses. Although such studies contribute significantly to the advancement of knowledge in the field, there is a gap in the literature regarding the detailed thermal characterization of specific textile waste with a focus on biomedical applications.

In this context, the objective of this study is to evaluate the feasibility of reusing different textile waste materials such as linen, denim, poplin, and helanca for use in orthopedic prostheses, investigating whether these materials have thermal and structural properties suitable for this purpose. To this end, thermal analyses were performed by thermogravimetry (TGA) and differential scanning calorimetry (DSC), in addition to determining density via pycnometry, seeking to understand the behavior of these residual fabrics and their compatibility with orthopedic device manufacturing processes. This study contributes original data to a little-explored area, clarifying the potential of these textile residues in the manufacture of eco-friendly prostheses.

Materials and Methods

Materials

In this study, four different types of textile waste were used in their raw form: linen, denim, poplin, and helanca, coded by the letters L, J, P, and H, respectively. The materials came from scraps discarded by a clothing factory located in the municipality of Campos dos Goytacazes, in the state of Rio de Janeiro.

After collection, the fabrics underwent a manual sorting process based on their composition and physical condition. They were then cut into standard sizes and kept in a dry environment until ready for use. Figure 1 illustrates the steps involved in obtaining and preparing these waste materials.

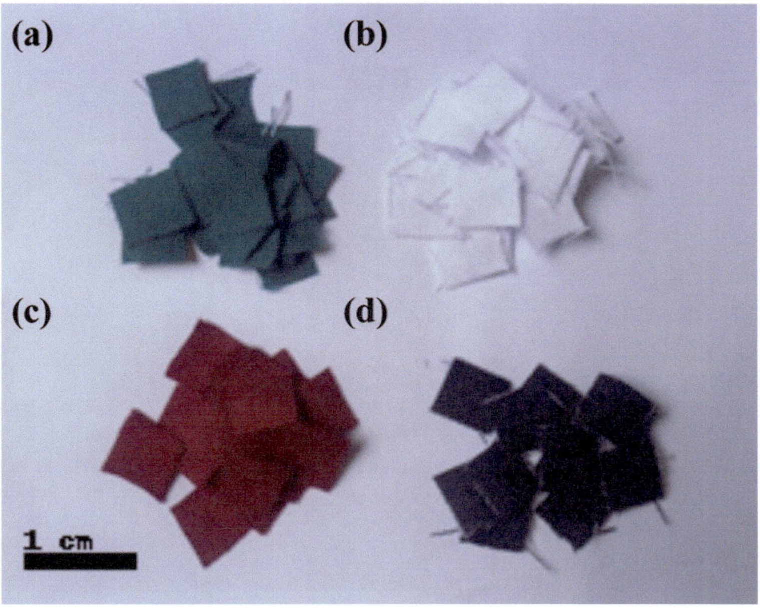

Fig. 1 Small fabric samples for analysis: **a** poplin, **b** linen, **c** helanca, **d** denim

Methodology—Density by Pycnometry

The density of linen, denim, poplin, and helanca fabrics was determined by pycnometry, following the guidelines of ASTM D792 (2020) [12]. Equation 1 was used, where (d) represents the density of the sample, (A1) is the mass of the empty pycnometer, (A2) is the mass of the pycnometer with the analyzed sample, (A3) is the mass of the pycnometer with the sample and water as shown in Fig. 2, and (A4) is the mass of the pycnometer with water only.

$$d = \frac{\mathbf{A2} - A1}{(A4 + A2) - (A1 + A3)} \tag{1}$$

where:

d = Density of the material;
A1 = mass of the empty pycnometer;
A2 = mass of the pycnometer + sample;
A3 = mass of the pycnometer + sample + water;
A4 = mass of the pycnometer + water.

Fig. 2 Pycnometer with poplin fabric being weighed

Liquid pycnometry density seeks to determine the actual density (or specific gravity) of a solid sample by measuring the volume of liquid displaced when the sample is immersed in a liquid of known density, usually water [13].

Methodology—Differential Scanning Calorimetry (DSC)

The samples, weighing between 5 and 10 mg, were placed in aluminum crucibles and subjected to thermal analysis under a nitrogen atmosphere, heated to 650 °C at a rate of 10 °C/min. The DSC curve was obtained using SDT650 thermogravimetric equipment from TA Instruments, located at SEPOL/LAMAV/UENF.

Methodology—Thermogravimetric Analysis (TGA)

Thermogravimetric analysis (TGA) was performed using SDT650 thermogravimetric equipment from TA Instruments, located at SEPOL/LAMAV/UENF. The tests were conducted in an inert nitrogen atmosphere, with a controlled heating rate of 10 °C/min, covering a temperature range of 650 °C. The samples analyzed, consisting of textile waste from linen, denim, poplin, and helanca fabrics in their pure form, were previously cut into small pieces to ensure greater uniformity during heating. They were then placed in platinum crucibles, suitable for high temperatures, for thermogravimetric analysis (TGA).

Results and Discussion

Table 1 shows the results of the densities of the tissues in their raw form.

Based on the data obtained by liquid pycnometry, the analyzed fabrics presented varying densities: linen (1.46 g/mL), helanca (1.47 g/mL), denim (0.90 g/mL), and poplin (0.34 g/mL), as shown in Table 1. The higher values for linen and helanca reflect the structural compaction and intrinsic density of the fibers that compose them, close to 1.45–1.50 g/cm^3. These results are in agreement with the study by

Table 1 Fabric density

Type of fabric	Density (g/ml)
Jeans	0.90
Poplin	0.34
Linen	1.46
Helanca	1.47

Rout et al. [14], which presents a model for estimating the actual density of fabrics from surface density and thickness, considering the effects of porosity and reinforcing the importance of considering porosity and thickness when estimating fabric density.

Figure 3a shows the DSC curve of helanca fabric, revealing thermal behavior typical of synthetic polymers, with moisture loss around 80 °C and the onset of thermal decomposition at 320 °C, related to the breakdown of polymer chains. Between 400 °C and 600 °C, an exothermic peak associated with oxidation and carbonization is observed. This thermal profile is compatible with the data presented by TG et al. [15], who reported similar behavior in hybrid fiber-reinforced epoxy composites, suggesting that materials with this performance can be considered for possible application in composites for orthopedic prostheses [16].

Figure 3b represents the DSC curve of the denim fabric and shows an endothermic peak around 80 °C, associated with moisture loss, followed by thermal stability up to about 320 °C, when the thermal decomposition of the cellulose fibers begins. A sharp exothermic peak near 390 °C indicates the degradation of organic chains and the onset of carbonization, a process that intensifies above 400 °C. This behavior demonstrates the thermal viability of denim for use as reinforcement in polymer composites, since, according to Baccouch et al. [17], denim fabric has good thermal and mechanical stability, withstanding the processing temperatures of epoxy composites.

The thermal transitions of jeans observed in DSC are consistent with the study by Isola et al. (2024), which highlighted the impact of chemical modifications on cotton. Silanization improves thermal stability, while sulfation-phosphorylation reduces the decomposition temperature and increases carbonization. Even without functionalization, jeans exhibit thermal behavior similar to that of cotton, indicating the potential of these treatments for technical applications [18].

Figure 3c shows the DSC curve of linen fabric, which presents an endothermic peak close to 90 °C, indicating moisture loss, and the onset of thermal decomposition around 300 °C, related to the degradation of cellulose present in natural fibers. An exothermic peak between 420 °C and 600 °C confirms the carbonization of the material. These results demonstrate that linen has good thermal stability up to 300 °C [19]. Linen, as a natural fiber with high strength and good thermal stability, stands out among the natural fiber-reinforced composites (NFRCs) analyzed by Castro-Franco et al. [22], being a promising candidate for application in orthopedic prostheses. According to the authors Oleiwi et al. [16], its mechanical and sustainable properties make flax suitable for composites that require lightness, structural performance, and biomechanical compatibility. The authors Neto et al. [18] point out that linen fabric has good thermal stability, with decomposition beginning at around 300 °C, suitable for composites processed at high temperatures, and that natural fibers with stability above 250 °C are considered suitable for this use.

The DSC curve of flax presented by Bahrami et al. [20] shows an endothermic peak between 40–150 °C, associated with moisture loss. Between 300–420 °C, thermal events related to the degradation of hemicellulose, cellulose, and lignin occur. These results confirm the typical thermal profile of lignocellulosic fibers.

Figure 3d, the DSC curve of poplin fabric shows a thermal profile compatible with fabrics composed of cotton and polyester, with thermal stability up to approximately

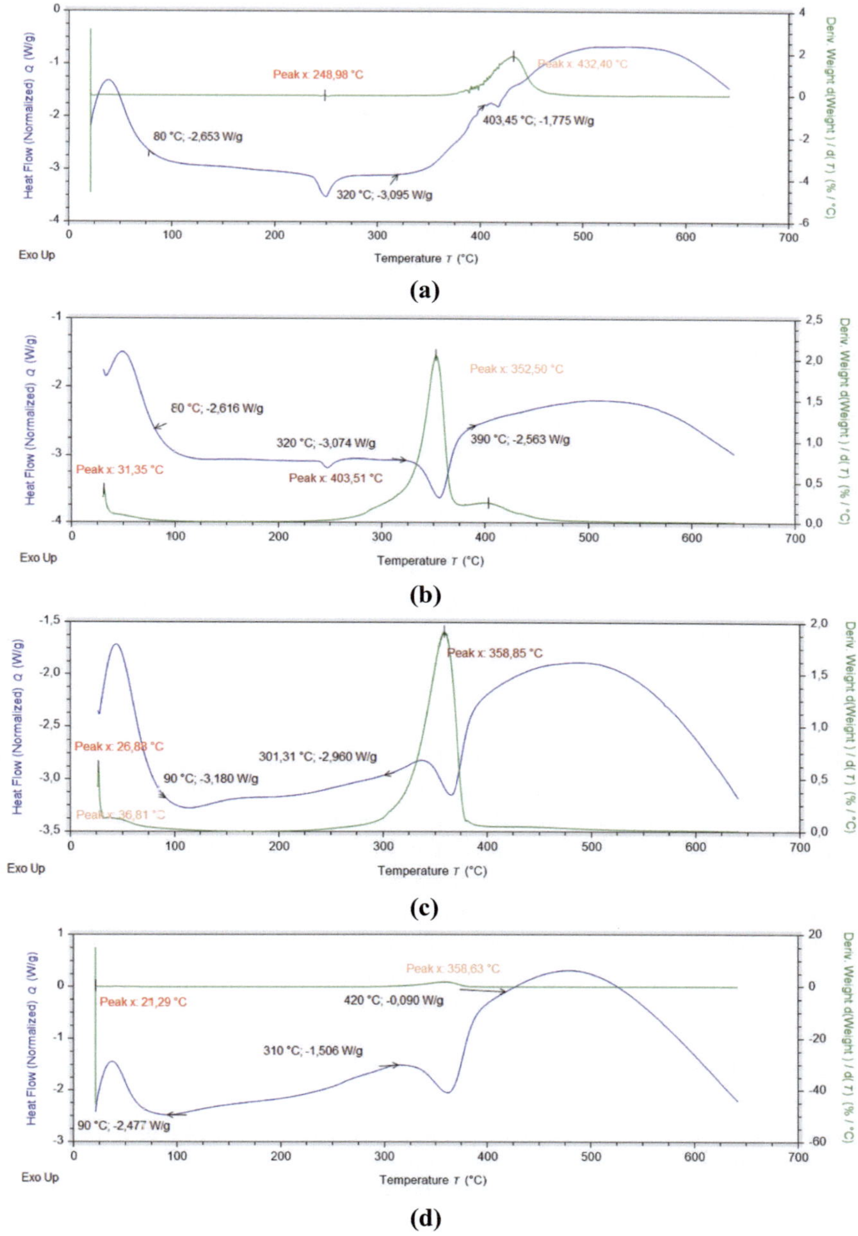

Fig. 3 DSC curves of fabrics: **a** helanca, **b** denim, **c** linen, and **d** poplin

300 °C and degradation beginning at 310 °C. Between 420 °C and 600 °C, the carbonization phase is observed, a behavior similar to that reported by Pereira et al. [21], who analyzed blended fabrics by TGA/DSC. The mixed fabrics analyzed in the study include combinations of polyester with cotton, polyester with wool, and ternary mixtures of polyester, cotton, and acrylic. These data suggest the thermal viability of poplin as a reinforcement material in composites for the manufacture of orthopedic prostheses [22].

The TGA analysis of helanca fabric, illustrated in figure A, shows thermal stability up to approximately 350 °C, followed by rapid decomposition up to 480 °C, with a residue of approximately 20%, associated with inorganic fractions or residual carbon. These results are in line with the study by Adler et al. [23], who investigated the thermal stability of recycled carbon fibers in epoxy composites, where an increase in degradation temperature and greater residual mass were observed compared to virgin fibers. This confirms that the presence of synthetic fibers, such as in Helanca, contributes to improved thermal stability and greater residue after decomposition (Fig. 4).

The TGA curve of denim fabric shows that the material maintains thermal stability up to approximately 290 °C, beginning to decompose in the range of 300–400 °C, with a final thermal residue of approximately 20% at 480 °C. These data indicate the predominant degradation of cellulose and the permanence of residual carbon, in agreement with the results obtained by Kamble et al. [24], who analyzed composites reinforced with recycled cotton fibers, including denim, and identified thermal decomposition between 250–450 °C, in addition to greater residual carbon generation.

TGA analysis of linen fabric indicated that the material retains its thermal stability up to approximately 280 °C, beginning its thermal decomposition in the range of 290 °C to 390 °C, a process related to the degradation of lignocellulosic fibers, as reported by Asim et al. [25] and Kumar et al. [26]. Mass loss reached approximately 80%, stabilizing with a residue of around 15% from 450 °C, which indicates the presence of residual carbon and inorganic components, as observed by Rahman et al. [27]. This thermal performance reinforces the potential of flax as a reinforcement in polymer composites for the manufacture of orthopedic prostheses, presenting adequate stability for the forming processes and functional requirements in biomedical applications, as discussed by Malalli et al. [28]. These findings support the use of natural fibers in environmentally sustainable composites, with performance compatible with biomedical engineering requirements.

Thermogravimetric analysis (TGA) of poplin fabric revealed thermal stability up to approximately 300 °C, with decomposition between 310 °C and 420 °C, attributed to the degradation of cellulosic and synthetic fibers, such as cotton and polyester, as reported by Abidnejad et al. [5] and Valente et al. [29]. Mass loss exceeded 85%, with a final residue of around 10% from 500 °C, indicating the presence of residual carbon or inorganic additives, in agreement with Spyridis et al. [30].

The observed thermal behavior confirms the potential of poplin fabric as reinforcement in polymer composites, especially due to its compatibility with thermal

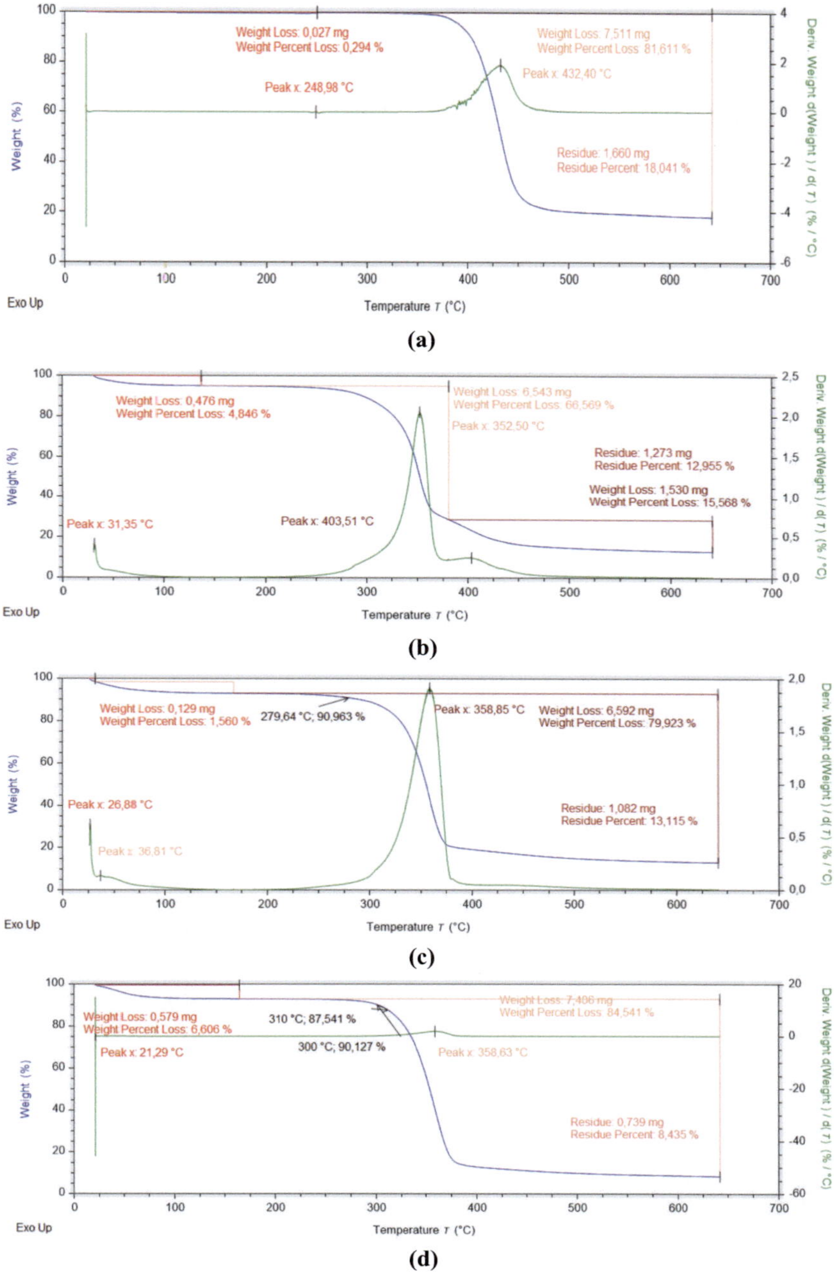

Fig. 4 TGA curves of fabrics: **a** helanca, **b** denim, **c** linen, and **d** poplin

molding processes. Studies by Patel et al. [31] reinforce this perspective, demonstrating that composites reinforced with natural fibers such as cotton, linen, and sisal, similar in composition to poplin exhibit good thermal stability, mechanical strength, and viability for applications in various areas.

Conclusion

The results obtained confirm that textile waste from linen, denim, poplin, and helanca have physical and thermal properties compatible with the processing and performance requirements of composites intended for low-cost, eco-friendly orthopedic prostheses. The higher density of linen (1.46 g/mL) and helanca (1.47 g/mL) indicates greater structural compaction, favoring mechanical integrity, while the lower density of denim (0.90 g/mL) and poplin (0.34 g/mL) suggests greater porosity, potentially beneficial for adhesion to the polymer matrix. DSC and TGA analyses showed adequate thermal stability between 280–350 °C, especially for helanca and linen, meeting the requirements for polymer matrix shaping and processing.

Based on the literature and experimental results, it appears that all fabrics analyzed have potential for application as reinforcement in biomedical composites, with helanca and poplin, due to their hybrid composition, offering thermal versatility and stability, while linen and denim, as cellulosic fibers, combine good thermal performance with sustainable attributes. Thus, these textile waste materials represent promising alternatives for the development of functional, durable, and environmentally responsible orthopedic prostheses, aligning with the goal of transforming industrial waste into value-added inputs, provided that their effectiveness is corroborated by future mechanical tests and clinical evaluations.

References

1. Sanjrani MA, Gang X, Mirza SNA (2025) A review on textile solid waste management: disposal and recycling. Waste Manag Res 43(4):522–529
2. Kumar S, Sahoo S (2021) Eco-sustainability of the textile production: waste recovery and current recycling in the composites world. Dogo Rangsang Res J 14(4):1916–1919
3. Bianco I et al (2023) Environmental impacts in the textile sector: a life cycle assessment case study of a woolen undershirt. Sustainability 15(15):11666
4. Juanga-LAbayen JP, Labayen IV, Yuan Q (2022) A review on textile recycling practices and challenges. Textiles 2(1):174–188
5. Abidnejad R et al (2025) High-fiber content composites produced from mixed textile waste: balancing cotton and polyester fibers for improved composite performance. Int J Biol Macromol 292:139227
6. Ghosh J et al (2025) Chemical valorization of textile waste: advancing sustainable recycling for a circular economy. ACS Omega 10(12):11697–11722
7. Nagarajan YR et al (2024) Self-reinforced composite materials: frictional analysis and its implications for prosthetic socket design. Materials 17(22):5629

8. Nagarajan YR et al (2024) Single polymer composites: an innovative solution for lower limb prosthetic sockets. Prosthesis. 6(3):457–477
9. Alimi L et al (2020) Structure and mechanical properties of PMMA/GF/Perlon composite for orthopedic prostheses. Mater Today Proc 31:S162–S167
10. Isola M et al (2024) Thermal properties and decomposition products of modified cotton fibers by TGA, DSC, and Py–GC/MS. Polym Degrad Stab 228:110937
11. Kumar S, Bhowmik S, Mahakur VK (2023) Thermo-mechanical and degradation properties of naturally derived biocomposites for prosthesis applications: analysis of the interface pressure and stress distribution on the developed socket. Proc Inst Mech Eng Part E J Process Mech Eng 237(5):1999–2015
12. ASTM (2020) Standard test methods for density and specific gravity (relative density) of plastics by displacement D792-20. West Conshohocken (PA): ASTM Int
13. Agrawal SA (2021) Simplified measurement of density of irregular shaped composites material using Archimedes principle by mixing two fluids having different densities. Int Res J Eng Technol 8(3):1005–1009
14. Rout PK, Singh MK (2023) Porosity determination of textile fabrics: a novel mathematical approach and experimental validation. Mater Today Commun 37:107559
15. Yg TG et al (2022) Mechanical and thermal properties of flax/carbon/kevlar based epoxy hybrid composites. Polym Compos 43(8):5649–5662
16. Oleiwi JK, Hamad QA, Faheed NK (2023) Experimental, theoretical, and numerical analysis of laminated composite prosthetic socket reinforced with flax and cotton fibers. Biotribology 35:100244
17. Baccouch W, Ghith A, Yalcin-Enis I, Sezgin H, Miiled W, Legrand X, Faten F (2022) Investigation of the mechanical, thermal, and acoustical behaviors of cotton, polyester, and cotton/polyester nonwoven wastes reinforced epoxy composites. J Ind Text 51(6):876–899
18. Neto JSS et al (2021) A review on the thermal characterisation of natural and hybrid fiber composites. Polymers 13(24):4425
19. Slapnik J, Lucyshyn T, Pinter G (2021) Relationships between the decomposition behaviour of renewable fibres and their reinforcing effect in composites processed at high temperatures. Polymers 13(24):4448
20. Bahrami M et al (2024) Thermal stability, durability, and service life estimation of woven flax-carbon hybrid polyamide biocomposites. Materials 17(9):2020
21. Pereira M et al (2024) Thermogravimetry as a tool to evaluate the contribution of different fabric constituents to fibre release. J Therm Anal Calorim 149(19):10497–10509
22. Castro-Franco AD et al (2024) A review of natural fiber-reinforced composites for lower-limb prosthetic designs. Polymers 16(9):1293
23. Adler SE et al (2021) Evaluation of recycled carbon fibre/epoxy composites: thermal degradation behaviour of pyrolysed and virgin carbon fibres using thermogravimetric analysis. Adv Ind Eng Polymer Res 4(2):82–92
24. Kamble Z et al (2022) Development and characterization of thermoset nanocomposites reinforced with cotton fibres recovered from textile waste. J Ind Text 51(2_suppl):2026S-2052S
25. Asim M et al (2020) Thermal stability of natural fibers and their polymer composites. Iranian Polym J 29(7):625–648
26. Kumar S et al (2024) Thermogravimetric analysis of lignocellulosic leaf-based fiber-reinforced thermosets polymer composites: an overview. Biomass Convers Biorefinery 14(12):12673–12698
27. Rahman MM, Karacan I (2022) The impact of eco-friendly chemical incorporation on the thermal oxidation process of flax fiber prior to carbonization and activation. J Mater Sci 57(3):2318–2333
28. Malalli CS, Ramji BR (2022) Mechanical characterization of natural fiber reinforced polymer composites and their application in prosthesis: a review. Mater Today Proc 62:3435–3443
29. Valente M, Rossitti I, Sambucci M (2023) Different production processes for thermoplastic composite materials: sustainability versus mechanical properties and processes parameter. Polymers 15(1):242

30. Spyridis Y et al (2024) Autonomous AI-enabled industrial sorting pipeline for advanced textile recycling. In: 2024 20th International conference on distributed computing in smart systems and the internet of things (DCOSS-IoT). IEEE, pp 455–461
31. Patel RV, Yadav A, Winczek J (2023) Physical, mechanical, and thermal properties of natural fiber-reinforced epoxy composites for construction and automotive applications. Appl Sci 13(8):5126

Mechanical Performance of a UV-Curable Resin Modified with Titanium Dioxide for SLA/LCD 3D Printing

Victor Paes Dias Gonçalves, Ellen Bernardes de Lima, Thayna Pereira Rangel, Noam Tonini Simonassi, Carlos Maurício Fontes Vieira, Henry Colorado Lopera, and Felipe Lopes Perisse

Abstract The advancement of additive manufacturing has driven the search for improved materials. Among these, photocurable resins stand out for their versatility and precision. However, they often exhibit fragile mechanical behavior and limited durability. This study aims to investigate the effect of titanium dioxide (TiO_2) reinforcement on the properties of a UV-curable acrylate resin used in SLA/LCD 3D printers. Samples were prepared by incorporating different concentrations of TiO_2 (0.5, 1.0%) into the base resin. Mechanical characterization was conducted following an accelerated aging protocol. It was observed that the 1.5% concentration compromised printability, preventing successful fabrication. Meanwhile, the 0.5 and 1.0% concentrations influenced the mechanical properties of the printed parts, with varying degrees of improvement. The results suggest that titanium dioxide can serve as a functional additive for enhancing specific properties in 3D-printed components, but only at low concentrations to maintain processability.

Keywords Photocurable resin · Titanium dioxide · Additive manufacturing

V. P. D. Gonçalves (✉) · E. B. de Lima · N. T. Simonassi · C. M. F. Vieira · F. L. Perisse
Advanced Materials Laboratory—LAMAV, State University of the Northern Rio de Janeiro—UENF, Campos dos Goytacazes, RJ, Brazil
e-mail: victor_paesdias@yahoo.com.br

T. P. Rangel
Fluminense Federal Institute of Education, Science and Technology (IFF Fluminense), Campos dos Goytacazes, RJ, Brazil

H. C. Lopera
CCComposites Laboratory, Engineering School, Universidad de Antioquia (UdeA), Medellín, Colombia

H. A. C. Lopera et al. (eds.), *Natural Fibers and Biocomposites 2026*, The Minerals, Metals & Materials Series, https://doi.org/10.1007/978-3-032-13577-3_11

Introduction

Polymeric materials have been gaining increasing attention in both research and industry, among them, resins stand out due to their versatility [1]. In the case of those used in stereolithography (SLA) 3D printing and in dentistry, they are easy to handle, adaptable to different applications, and allow for property adjustments according to specific needs [2, 3]. Despite these advantages, there are still aspects that require improvement, enhancing mechanical strength, durability, and thermal stability without compromising aesthetics or biocompatibility remains a challenge [3, 4].

One promising approach is the incorporation of inorganic nanoparticles; titanium dioxide (TiO_2) is one of the most notable candidates, as it combines several desirable characteristics: high mechanical strength, chemical stability, a high refractive index, and photocatalytic activity [5]. Moreover, it is already recognized as biocompatible, which facilitates its application in medical and dental fields [6].

In dentistry, TiO_2-reinforced composites demonstrate greater wear resistance and reduced biofilm formation due to the antibacterial effect of the particles [7]. In 3D printing, the addition of TiO_2 can improve impact resistance, enhance dimensional stability, and provide a more uniform and durable surface finish [6].

However, excessive amounts of TiO_2 can lead to nanoparticle agglomeration, which results in printing defects and may even compromise the mechanical properties of the material [7]. Therefore, determining the optimal concentration and ensuring proper nanoparticle dispersion are crucial. The objective of this study is to investigate how different concentrations of TiO_2 affect polymeric resins produced by 3D printing. The aim is to evaluate changes in their mechanical, thermal, and structural properties and, based on these findings, discuss the potential applications of these resins in both industrial and biomedical fields.

Methodology

Material and Preparation

Specimens were prepared using a stereolithography-based LCD 3D printer Creality HALOT-MAGE 8K (Fig. 1) with a commercial Eco Resin as the base matrix. To enhance mechanical performance and potentially improve impact resistance, the resin was homogeneously blended with titanium dioxide (TiO_2) nanoparticles at a concentration of 1%.

The mixture was magnetically stirred and sonicated for 30 min to ensure uniform dispersion of the inorganic phase, thereby minimizing agglomeration and promoting consistent photopolymerization throughout the printed parts.

Printing was carried out with a layer thickness of 0.05 mm, employing six base layers cured for 50 s each, followed by subsequent layers with an exposure time of

Fig. 1 Creality HALOT-MAGE 8K

4 s. It was observed that the 1.5% concentration compromised printability, preventing successful fabrication.

Following fabrication (Fig. 2), the specimens underwent a two-step post-processing protocol: first, two washing cycles in isopropyl alcohol (2 × 5 min) were performed to remove uncured resin residues; second, samples were air-dried for 15 min. Final curing was achieved under controlled irradiation at 405 nm for 20 min to ensure full crosslinking of the resin matrix. The incorporation of TiO_2 was expected to influence both the curing kinetics and the fracture behavior of the polymer, providing a valuable framework to investigate the effect of inorganic fillers on the toughness of SLA/LCD-printed photopolymers.

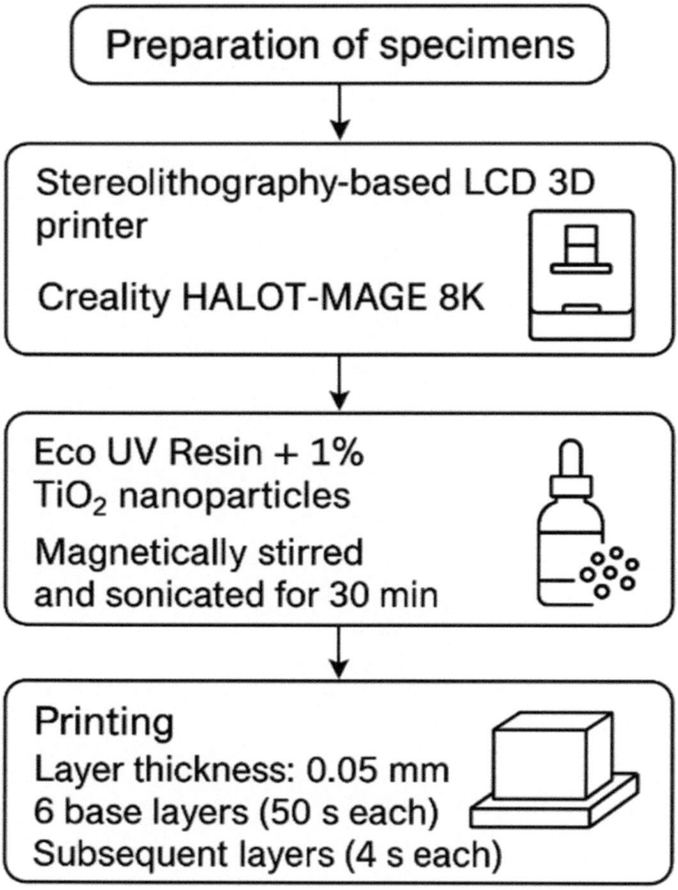

Fig. 2 Flowchart

Compression

Compression testing of the polymer samples was conducted to evaluate their mechanical behavior under uniaxial load. Cylindrical specimens were prepared according to ASTM D695 [8] standards. Tests were performed using a universal testing machine (Fig. 3) equipped with a 5 kN load cell at a crosshead speed of 1.3 mm/min at room temperature 23 ± 2 °C. Each sample was placed between the compression plates, ensuring proper alignment to avoid bending stresses. Eleven specimens per formulation were tested, and the mean values and standard deviations were reported.

Fig. 3 Instron 5582 universal testing machine

Impact Izod

The Izod impact test was carried out in accordance with ASTM D256 [9] to assess the dynamic fracture resistance of the polymeric specimens. Rectangular samples with dimensions of 63.5 mm × 12.7 mm × 3.2 mm and a standardized V-notch of 2.54 mm depth at 45° were prepared to ensure reproducibility of crack initiation. Each specimen was rigidly clamped in a vertical cantilever position, with the notched side facing the pendulum striker. A calibrated pendulum hammer with a capacity of 22 J was released from a predefined height, delivering a single impact to the specimen (Fig. 4). The absorbed energy was determined from the reduction in potential energy of the pendulum after fracture, normalized by specimen thickness, and expressed in J/m.

Fig. 4 PANTEC brand XC-50 impact machine

Shore D Hardness

Surface hardness was assessed in accordance with ASTM D2240 [10] using a calibrated Shore D durometer. Flat specimens with a minimum thickness of 6 mm were fabricated to eliminate substrate interference, and measurements were taken after 1 s of contact at no fewer than five distinct points, each separated by at least 6 mm. Edge regions and zones influenced by printing supports were excluded to avoid bias in the results.

Wettability (Contact Angle)

Surface wettability was characterized by static water contact angle measurements following ASTM D7334 [11]. Prior to testing, the specimens were carefully cleaned with isopropyl alcohol and air-dried, and analyses were performed 1 h after post-curing to ensure stabilization of surface chemistry. Droplets of deionized water (3–5 μL) were deposited on the flat surfaces and left and right contact angles were recorded within 5–10 s of droplet placement. Wettability is a critical parameter reflecting both surface energy and chemical functionality; alterations can occur due to residual unreacted monomers, degree of curing, or surface enrichment of TiO_2 particles.

Density

The bulk density of the printed specimens was determined according to ASTM D792 [12], using the buoyancy method in deionized water at controlled room temperature (23 ± 2 °C). This parameter is directly influenced by both the chemical composition and the structural integrity of the polymeric matrix.

Statistical Analysis (ANOVA)

Statistical analyses were performed using one-way analysis of variance (ANOVA) to evaluate the effect of experimental variables on the measured properties. This method was selected due to its robustness in determining whether statistically significant differences exist among the mean values of multiple groups while controlling for type I error. The use of ANOVA is particularly relevant in this study to compare the influence of TiO_2 incorporation and processing variables on mechanical and physicochemical outcomes, providing a rigorous basis for interpreting trends across experimental conditions.

Results and Discussion

Compression testing (Fig. 5) demonstrated that the 0% formulation exhibited an initial compressive strength of 160 ± 3.2 MPa, while the 1% TiO_2 formulation started at 132 ± 2.8 MPa, indicating that the addition of the filler reduced the material's baseline mechanical performance. After UV exposure, the compressive strength decreased in both formulations, with the 0% resin dropping to 130 ± 3.6 MPa and the 1% TiO_2 resin declining to 109 ± 5.9 MPa. These results indicate that the polymer matrix is inherently susceptible to mechanical degradation, and UV radiation exacerbates this effect, particularly in the TiO_2-containing formulation. Overall, the data suggest that the presence of TiO_2 at 1% does not provide protection against UV-induced weakening and may even contribute to a more pronounced reduction in compressive strength under aging conditions.

Izod impact testing (Fig. 6) revealed no significant differences in impact resistance among the tested formulations and conditions. Both the 0 and 1% TiO_2 resins exhibited similar values before and after UV exposure, indicating that neither the incorporation of TiO_2 nor UV-induced degradation affected the material's ability to absorb sudden impact energy. Importantly, the fracture type remained brittle in all cases, suggesting that despite the stability in measured impact values, the resin's failure mechanism is controlled by its inherent brittleness rather than by changes in formulation or UV aging.

Density analysis (Table 1) further confirmed the impact of UV-induced degradation. Both the pure resin and the TiO_2-modified formulation exhibited decreased density after UV exposure, suggesting mass loss and potential formation of microvoids within the polymer matrix. The presence of 1% TiO_2 did not prevent

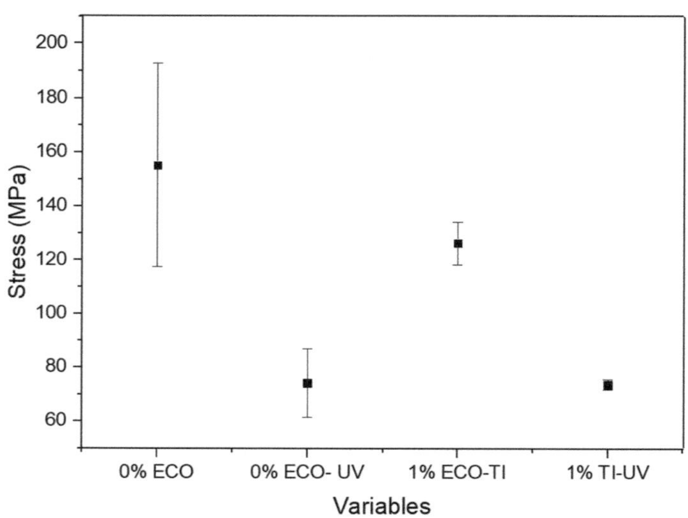

Fig. 5 Results compression stress

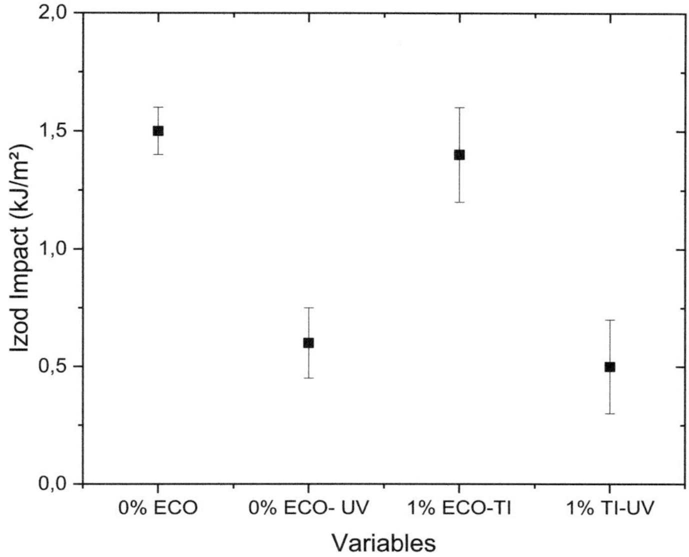

Fig. 6 IZOD impact

this effect, highlighting that at this concentration, the additive is not sufficient to mitigate structural degradation.

The Shore D hardness test (Table 2) showed that UV radiation was capable of significantly reducing the resin's hardness, indicating degradation of the polymer matrix after exposure. However, the addition of 1% TiO_2 did not provide any protective effect or notable change in the measured values, both before and after aging. This behavior suggests that, at the concentration used, titanium dioxide was insufficient to act as a stabilizer against UV radiation, leaving the material susceptible to the same alterations observed in the pure resin. Therefore, the observed degradation can be primarily attributed to the action of UV radiation, regardless of the presence of the additive.

In the wettability test (Table 3), no significant differences were observed among the evaluated conditions, indicating that neither UV radiation nor the addition of 1% titanium dioxide was able to meaningfully alter the contact angle of the 3D resin.

Table 1 Density results

Sample	Results (g/cm^3)	Statistical (vs. pre UV)
Resin 0% (pre UV)	1.185($\pm$0.004)	–
Resin 1% TiO_2 (pre UV)	1.192($\pm$0.003)	–
Resin 0% (post UV)	1.072($\pm$0.005)	$p < 0.05$
Resin 1% TiO_2 (post UV)	1.080($\pm$0.004)	$p < 0.05$

Table 2 Results shore D

Sample	Results (D)	Statistical (vs. pre UV)
Resin 0% (pre UV)	83(±1.5)	–
Resin 1% TiO_2 (pre UV)	82(±2.0)	–
Resin 0% (post UV)	76(±1.0)	$p < 0.05$
Resin 1% TiO_2 (post UV)	75(±2.0)	$p < 0.05$

This finding suggests that the polymer matrix maintains surface stability, preserving hydrophobic behavior regardless of the applied treatments. The lack of effect may be related to the low TiO_2 concentration used or to limited penetration and activation of UV radiation at the material surface.

These two tests were performed complementarily to evaluate distinct aspects of the 3D resin's stability under UV exposure and TiO_2 incorporation. While the Shore D hardness test assesses changes in the mechanical strength of the polymer matrix, revealing structural degradation, the wettability test investigates changes in surface energy, reflecting possible chemical or topographical modifications that could affect liquid interaction.

Hafkamp et al. [13] and Ryu et al. [14] said that ceramic additives, while potentially beneficial for modifying material properties, present notable challenges in photopolymerizable resins. Their incorporation increases the viscosity of the resin, which can compromise the flow behavior and affect the printing process. Additionally, internal light scattering caused by the ceramic particles reduces the effective curing depth, potentially decreasing the resolution and surface quality of printed parts. Therefore, parameters such as particle size and size distribution, ceramic powder concentration, and the homogeneity of the final resin-powder mixture must be carefully optimized to mitigate these drawbacks and ensure consistent mechanical performance and dimensional accuracy.

Table 3 Results of wettability

Sample	Contact angle	Classification	Statistical
Resin 0% (pre UV)	72.5(±2.3)	Hydrophilic	=0.86
Resin 1% TiO_2 (pre UV)	70.8(±1.9)	Hydrophilic	
Resin 0% (post UV)	75.4(±2.1)	Hydrophilic	
Resin 1% TiO_2 (post UV)	71.7(±4,5)	Hydrophilic	

Conclusion

The addition of TiO_2 powder to the resin posed challenges in the photopolymerization process using LED light. The presence of the powder can scatter or absorb the incident light, reducing the penetration depth and intensity of the LED radiation within the resin matrix. This can lead to incomplete curing, particularly in thicker regions, and may compromise the formation of a fully crosslinked network. As a result, even low concentrations of TiO_2 can affect the mechanical properties and surface characteristics of the printed parts, highlighting the need for careful optimization of filler content when using LED-based curing systems.

References

1. Yüceer ÖM, Kaynak Öztürk E, Çiçek ES, Aktaş N, Bankoğlu Güngör M (2025) Three-dimensional-printed photopolymer resin materials: a narrative review on their production techniques and applications in dentistry. Polymers 17:316. https://doi.org/10.3390/polym17030316
2. Stajcic I, Veljkovic F, Petrovic M, Veličkovic S, Radojevic V, Vlahović B, Stajcic A (2023) Impact- and thermal-resistant epoxy resin toughened with acacia honey. Polymers 15:2261. https://doi.org/10.3390/polym15102261
3. Jeong M, Radomski K, Lopez D, Liu JT, Lee JD, Lee SJ (2024) Materials and applications of 3D printing technology in dentistry: an overview. Dent J 12:1. https://doi.org/10.3390/dj12010001
4. Dias Gonçalves VP, Vieira CMF, Simonassi NT, Perissé Duarte Lopes F, Youssef G, Colorado HA (2024) Evaluation of mechanical properties of ABS-like resin for stereolithography versus ABS for fused deposition modeling in three-dimensional printing applications for odontology. Polymers 16:2921. https://doi.org/10.3390/polym16202921
5. Mubarak S, Dhamodharan D, Divakaran N, Kale MB, Senthil T, Wu L, Wang J (2020) Enhanced mechanical and thermal properties of stereolithography 3D printed structures by the effects of incorporated controllably annealed anatase TiO_2 nanoparticles. Nanomaterials 10:79. https://doi.org/10.3390/nano10010079
6. Altarazi A, Haider J, Alhotan A, Silikas N, Devlin H (2023) 3D printed denture base material: the effect of incorporating TiO_2 nanoparticles and artificial ageing on the physical and mechanical properties. Dent Mater 39(12):1122–1136. https://doi.org/10.1016/j.dental.2023.10.005. Epub 2023 Oct 13. PMID: 37839997
7. Chen S-G, Yang J, Jia Y-G, Lu B, Ren L (2019) TiO_2 and PEEK reinforced 3D printing PMMA composite resin for dental denture base applications. Nanomaterials 9:1049. https://doi.org/10.3390/nano9071049
8. ASTM D256-10 (2010) Standard test methods for determining the izod pendulum impact resistance of plastics. ASTM International, West Conshohocken
9. ASTM D695-15 (2015) Standard test method for compressive properties of rigid plastics. ASTM International, West Conshohocken
10. ASTM D2240-15 (2015) Standard test method for rubber property—durometer hardness. ASTM International, West Conshohocken
11. ASTM D7334-08 (2008) Standard practice for surface wettability of coatings, substrates and pigments by advancing contact angle measurement. ASTM International, West Conshohocken
12. ASTM D792-20 (2020) Standard test methods for density and specific gravity (relative density) of plastics by displacement. ASTM International, West Conshohocken

13. Hafkamp T, van Baars G, de Jager B, Etman P (2018) A feasibility study on process monitoring and control in vat photopolymerization of ceramics. Mechatronics 56:220–241
14. Ryu K, Kim J, Choi J, Kim U (2022) The 3D printing behavior of photocurable ceramic/polymer composite slurries prepared with different particle sizes. Nanomaterials 12:2631

Part IV
Poster Session

A Review of Natural Bioactives as a Sustainable Way of Controlling Biofouling

Luiza Fortunato Freitas, Victor Paes Dias Gonçalves, Michel Picanço Oliveira, Noan Tonini Simonassi, Carlos Maurício Fontes Vieira, and Felipe Perissé Duarte Lopes

Abstract The colonization of submerged surfaces by marine organisms, a phenomenon known as biofouling, represents a significant challenge for aquatic structures, especially in naval and offshore environments. This process, which begins with the formation of microbial biofilms and evolves into the fixation of macrofouling, causes damage such as bioinvasion, increased hydrodynamic drag, high fuel consumption, and increased maintenance costs. The use of natural biocides, although effective, generates toxicity for non-target organisms and ecological impacts. In this view, this paper carries out a literature review on more sustainable alternatives for combating biofouling, focusing on substances extracted from natural sources that present a viable strategy for inhibiting the initial stages of biofouling, promoting the protection of aquatic structures with less ecological risk.

Keywords Bioactive · Sustainability · Biofouling · Corrosion

Introduction

Biofouling is a natural phenomenon characterized by the accumulation of organic matter and minerals on submerged surfaces. This process occurs in successive stages. Initially, a thin film forms on the metal surface due to the adsorption of proteins, glycoproteins, and polysaccharides. Next, marine microorganisms colonize this layer, forming a biofilm. Subsequently, diatoms, microalgae spores, and larvae adhere to the surface, stabilizing the encrusting community. Finally, macroorganisms, such as algae and invertebrates, develop on the surface, consolidating a

L. F. Freitas (✉) · V. P. D. Gonçalves · N. T. Simonassi · C. M. F. Vieira · F. P. D. Lopes
Advanced Materials Laboratory—LAMAV, State University of the Northern Rio de Janeiro—UENF, Campos dos Goytacazes, RJ, Brazil
e-mail: 20251200034@pq.uenf.br

M. P. Oliveira
Department of Forestry and Wood Sciences, Federal University of Espírito Santo, Jeronimo Monteiro, ES, Brazil

H. A. C. Lopera et al. (eds.), *Natural Fibers and Biocomposites 2026*, The Minerals, Metals & Materials Series, https://doi.org/10.1007/978-3-032-13577-3_12

complex ecosystem [1]. These stages are represented in Fig. 1, which schematically illustrates the successive development of biofouling.

This phenomenon has serious economic and environmental consequences. One of the most obvious impacts is the increase in maintenance costs for submerged structures, which require frequent cleaning, repainting, and repairs. In addition, the formation of biofilms promotes biocorrosion, a process in which substances secreted by microorganisms accelerate the degradation of metal surfaces, as shown in Fig. 2.

A comparison between abiotic and biotic environments shows that the presence of microorganisms intensifies the degradation of metal surfaces [3].

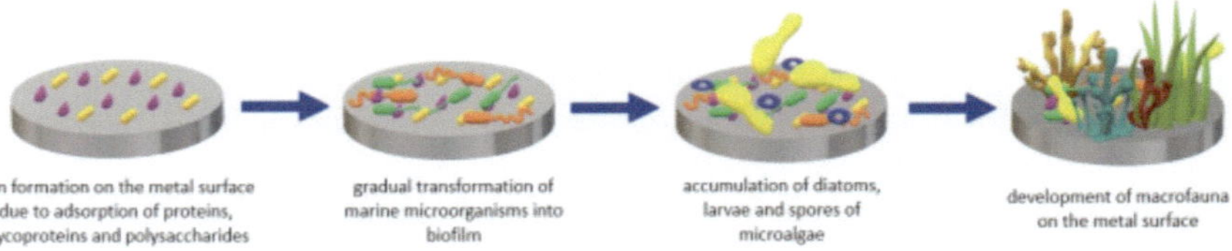

Fig. 1 Process of marine biofouling [2]

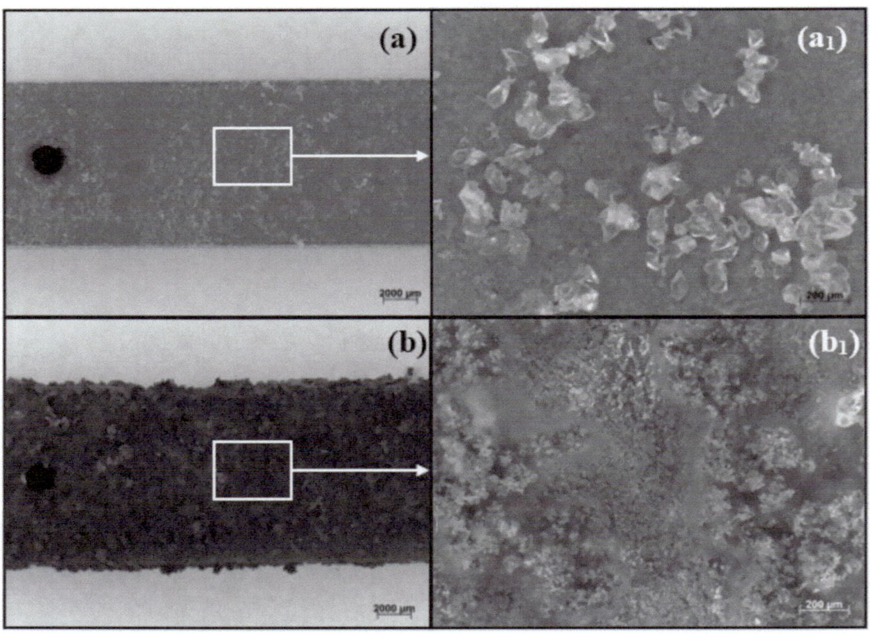

Fig. 2 Macrographs of the film and biofilm formed on the X80 pipeline steel surface after 15 days of exposure to the abiotic (**a** and $\mathbf{a_1}$) and biotic (**b** and $\mathbf{b_1}$) systems, respectively. Magnification of 7× (**a** and **b**) and 100× ($\mathbf{a_1}$ and $\mathbf{b_1}$) [3]

Another significant impact is bioinvasion: organisms encrusted on ship hulls can be transported to different regions and, in the absence of natural predators, establish themselves as invasive species. A notable example is the invasive coral *Tubastraea* spp. (sun coral), which has spread along the Brazilian coast, competing with native species and threatening local biodiversity (Fig. 3).

In this context, this study aims to conduct a systematic review of the literature on the use of natural bioactive compounds as sustainable alternatives for controlling marine biofouling. To this end, a bibliographic search was conducted in scientific databases, selecting articles that address the identification, origin, and application of these compounds in coatings or laboratory tests. Thus, this review discusses the impacts of biofouling, the limitations of synthetic biocides, and the prospects for the use of natural substances in the development of strategies that are less harmful to the environment.

Methodology

The literature search was conducted in the Scopus, ScienceDirect, and PubMed databases, supplemented by searches in Google Scholar and SpringerLink, to ensure comprehensive coverage of the available literature on the subject. Subsequently, the data extracted from Scopus were imported into the Bibliometrix package (R environment) for refinement and bibliometric analysis.

To develop the search strategy, descriptors in Portuguese and English were defined related to the central theme of the research: marine biofouling and the use of natural bioactive compounds as sustainable alternatives for biofouling control. These descriptors were identified by consulting technical terms and thesauros used in scientific literature. Table 1 presents the set of terms used, organized by thematic categories (biofouling, biocorrosion, and bioactive).

Boolean operators were used in the search strategy, a feature that allows the logical combination of keywords to refine the results obtained in the databases:

- **OR**: used to cover synonyms or related terms within the same category. For example, "biofouling OR bioincrustation" broadens the search to include articles containing either of the two terms. This operator is essential for increasing the sensitivity of the search, ensuring that no relevant articles are excluded due to terminological variation.
- **AND**: used to relate different categories of terms, restricting the search to articles that contain both keywords simultaneously. For example, "biofouling AND biocorrosion AND bioactive" returns only studies that address all three aspects, making the search more specific and aligned with the objective of the review.

Thus, the combination of OR (for synonyms within each group) and AND (for crossing between different groups) allowed us to construct robust and targeted search expressions, ensuring both breadth and specificity in the retrieval of articles.

Fig. 3 Invasive corals *Tubastraea* spp. **a** Living colony of *T. coccinea*; **b** living colony of *T. tagusensis*; **c** complete skeleton of colony of *T. coccinea*; **d** complete skeleton of colony of *T. tagusensis*; **e** detail of corallite and septal arrangement in *T. coccinea*; **f** detail of corallite and septal arrangement in *T. tagusensis*. Bars = 1 cm [6]

Table 1 Search strategy

A REVIEW OF NATURAL BIOACTIVES AS A SUSTAINABLE WAY OF CONTROLLING BIOFOULING			
Portuguese:	**Bioincrustação**	**Biocorrosão**	**Bioativo**
Terms:	Biofouling	Biocorrosion	Bioactive
English:	Infection	Decomposition	Biological
	Decomposition	Deterioration	Primitive
Thesauros:	-	Erosion	-
	-	Oxidation	-

OR (↓ down the rows) AND (→ across the columns)

Inclusion and Exclusion Criteria

Inclusion

- Publications between 2020 and 2025.
- Studies addressing natural bioactive compounds in biofouling control.
- Original articles and systematic reviews.
- Texts in English and Portuguese.

Exclusion

- Studies focused exclusively on synthetic or metallic biocides.
- Works outside the scope.
- Publications without access to the full text.

Bibliometrix Package

After extracting the data from the Scopus database using the search strategy described in Table 1, the information was imported into the Bibliometrix package (R environment) for a more detailed bibliometric analysis. This tool allowed for a refined filtering of the data, enabling the identification of publication trends, the most relevant journals, and the main contributing countries. Through this process, it was possible to obtain a clearer and more structured overview of the scientific production related to sustainable alternatives for combating biofouling and biocorrosion.

Analysis Tools

The articles included in the review were organized in a spreadsheet (Excel), from which information such as year of publication, type of natural compound, biological property, and form of application were extracted. This systematization allowed us to identify patterns and trends in the literature.

In spreadsheet, it can be seen that the articles analyzed concentrate on the period from 2022 to 2024, focusing on the investigation of natural compounds with biological properties and the development of more sustainable alternatives to combat marine biofouling.

Results

The application of the search strategy resulted in two graphs that illustrate the growing scientific interest in combating biofouling and biocorrosion through more sustainable alternatives, such as the bioactive compounds listed in Table 2. These graphs were generated using data from the Scopus database, covering the period from 2020 to 2025, based on the search strategy described in Table 1.

Figure 4 shows a noticeable increase in the number of publications related to biofouling and biocorrosion in 2021, followed by a decline in 2022. Nevertheless, between 2020 and 2025, a total of 80 articles on the topic were published, confirming the continuity of research interest in this area.

In addition, Figs. 7 and 8 were generated through Bibliometrix analysis, providing complementary insights into source productivity and country contributions, which reinforce the global scientific engagement with sustainable approaches to biofouling and biocorrosion control.

Table 2 Spreadsheet of bioactive compounds

Year	Natural compound	Biological properties	Application
2024	Natural flavonoids [7]	Antimicrobial Anti-inflammatory Antioxidant	Suggests incorporation of antifouling paints
2022	Analogs of bromosphaerol (isolated from red alga *Sphaerococcus coronopifolius*) [8]	Antimicrobial Antifungal	Bioinspired alternatives in antifouling coatings and paints
2023	Extracts of *Jania rubens* and *Bryothamnion seaforthii* from the phylum *Rhodophyta* [9]	Antifouling	Ecological paints
2023	Acacia catechu (catechin) [10]	Antifouling Antibacterial	Can be used as natural product antifouling agents

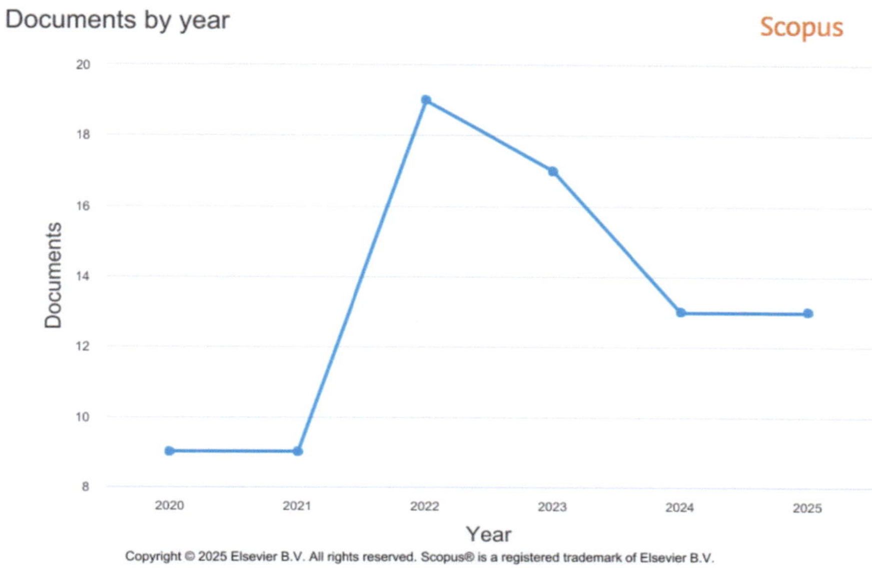

Fig. 4 Biofouling AND biocorrosion

Figure 5 shows the evolution of publications on biocorrosion and bioactives. In 2021, there was a significant increase, with seven articles published, representing the peak of interest in the topic. However, in 2022 and 2023, there was a significant drop, with only two and one document registered, respectively. From 2024 onwards, there is a gradual resumption of publications, reaching three articles in 2024 and five in 2025. These data suggest a cyclical movement in scientific interest, but they show a recent trend towards the appreciation of bioactives as sustainable strategies in combating biocorrosion.

Figure 6 shows the annual distribution of publications on biofouling and bioactive substances as control alternatives from 2020 to 2025. It can be seen that 17 articles were published in 2020, a number that fell to 12 in 2021. In 2022, there was a recovery, reaching 18 publications, followed by a slight drop in 2023 (14 articles). The year 2024 stood out as the year with the highest scientific production in the period, with 26 publications, representing the peak of interest in the topic. In 2025, however, there was a significant drop, with 11 records. This behavior highlights fluctuations in scientific engagement but reinforces that the topic has been attracting growing attention, especially in the context of sustainable solutions based on bioactives for biofouling control.

The data show that, despite fluctuations in the number of articles published, there is a growing trend in scientific interest in sustainable solutions based on bioactives, indicating the increasing relevance of this line of research for the control of biofouling and biocorrosion (Fig. 7).

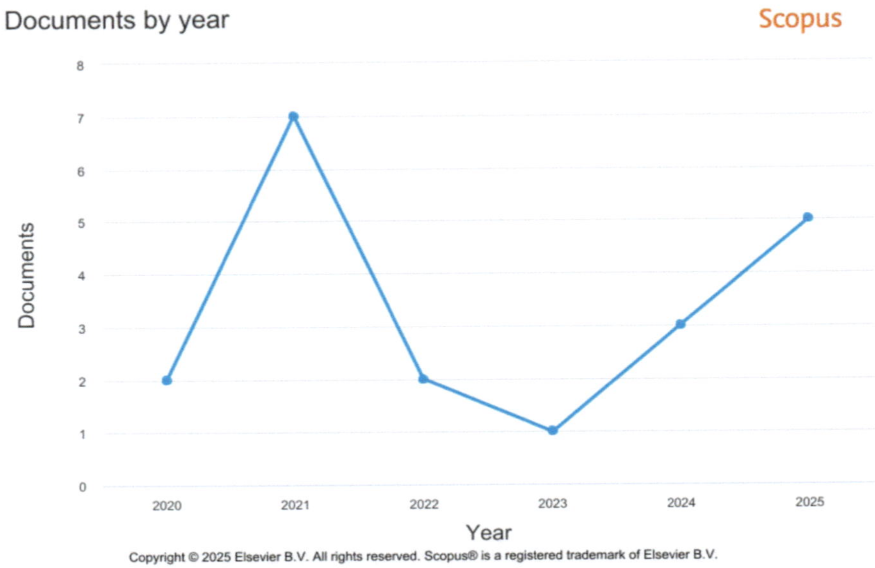

Fig. 5 Biocorrosion AND bioactive

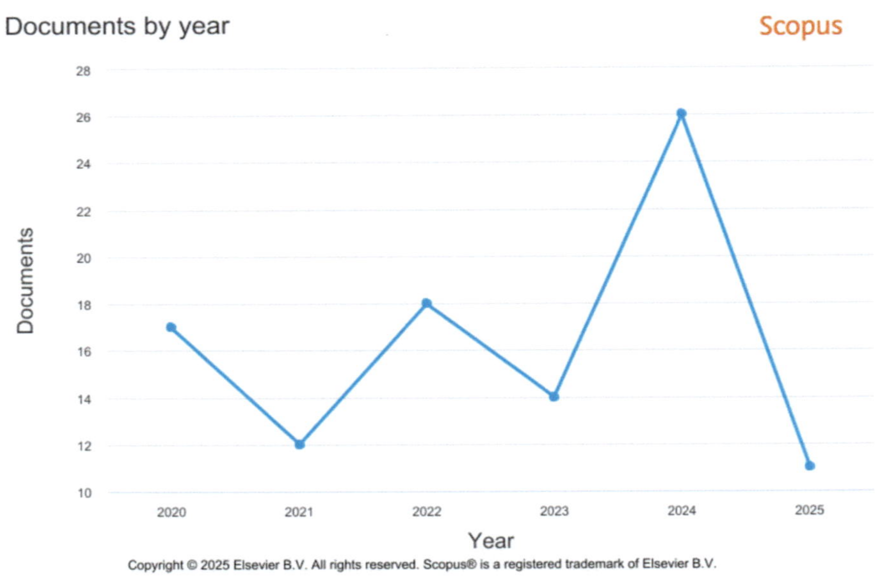

Fig. 6 Biofouling AND bioactive

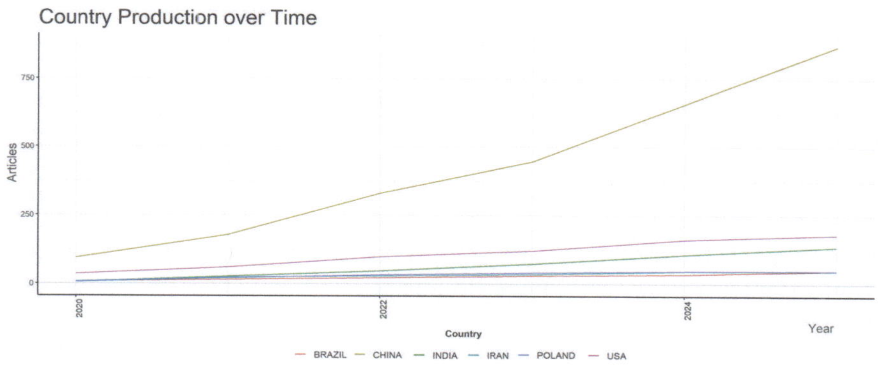

Fig. 7 Country production over time

The graph displays the evolution of the number of articles published by different countries from 2020 to 2025. China dominates scientific production with a sharp and continuous increase over the years. The United States ranks second, followed by India and Iran, both showing upward trends. Poland and Brazil contribute fewer publications but exhibit steady growth, highlighting the global strengthening of research efforts in this field (Fig. 8).

The graph illustrates the cumulative evolution of publications from the main journals between 2020 and 2025. *Chemosphere* stands out with the highest and most consistent growth, maintaining its position as the leading source. *Science of the Total Environment* and *Journal of Environmental Management* also show a steady increase in publication volume. Meanwhile, *Chemical Engineering Journal* and *Water Research* display moderate yet continuous growth, suggesting a gradual expansion of scientific output in these journals.

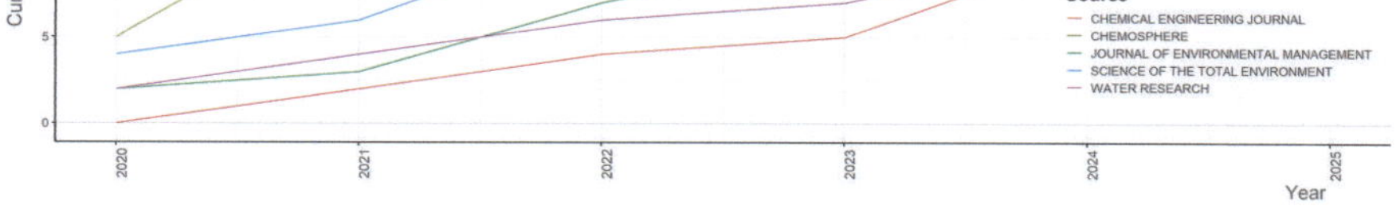

Fig. 8 Sources production over time

Conclusion

This study demonstrated that there are sustainable alternatives for combating biofouling, a phenomenon that triggers associated processes such as biocorrosion and bioinvasion. The literature review showed that naturally occurring bioactive compounds have significant potential to mitigate these impacts, acting mainly during the initial stages of biofilm development. Their mechanisms include the inhibition of microbial adhesion to submerged surfaces and the disruption of quorum sensing (QS), the cell-to-cell communication system that regulates bacterial cooperation, gene expression, and biofilm maturation. By interfering with QS signaling molecules, such as acyl-homoserine lactones (AHLs), these natural biomolecules can prevent microorganisms from reaching the population density required to coordinate biofilm formation. In addition to offering an effective strategy for controlling biofouling, these compounds present lower environmental risks compared to conventional synthetic biocides.

However, there is a need to expand field-scale studies and develop more stable and effective formulations in order to enable the practical application of these bioactives in naval and industrial contexts. Thus, the use of natural substances is a relevant and expanding prospect for the development of safer antifouling technologies that are aligned with the principles of sustainability.

Acknowledgements The authors thank the Brazilian agencies CNPq, process number 302976/2022-1, and FAPERJ, process number E-26/200.847/2021.

References

1. Xia Z, Gu J, Wen Y, Cao X, Gao Y, Li S et al (2024) EDNA-based detection reveals invasion risks of a biofouling bivalve in the world's largest water diversion project. Ecol Appl 34(1):e2826. https://doi.org/10.1002/eap.2826
2. Kubiak A, Kotula M, Leśniewski B, Pajewska-Szmyt M (2023) Iron-sponges interrelations: from biocorrosion to nanostructured biocomposites. Lett Interdiscip Nat Sci 12(3):64. https://doi.org/10.33263/LIANBS123.064
3. Oliveira ESD, Pereira RFC, Lima MAGA, Urtiga Filho SL (2021) Study on biofilm forming microorganisms associated with the biocorrosion of X80 pipeline steel in produced water from oilfield. Mater Res 24(6):e20210196. https://doi.org/10.1590/1980-5373-MR-2021-0196
4. Bonilla OH, Nascimento SF, Lucena EMP, Filho JVR, Pinheiro HB (2023) Distribuição e Intensidade de Invasão de Cryptostegia spp. em Quatro Estados no Nordeste do Brasil. Rev Bras Geogr Fís 16(2):986–1000. https://doi.org/10.26848/rbgf.v16.2.p986-1000
5. Kartsonakis IA, Saji VS, Tziveleka LA, Singh R, Blackwood DJ, Gu T (2023) Editorial: biofouling, biocorrosion and biodeterioration: recent advancements. Front Bioeng Biotechnol 11:1144671. https://doi.org/10.3389/fbioe.2023.1144671
6. Silva A, Lima R, Gomes A, Fleury B, Creed J (2011) Expansion of the invasive corals Tubastraea coccinea and Tubastraea tagusensis into the Tamoios Ecological Station marine protected area, Brazil. Aquat Invasions 6:S1.024. https://doi.org/10.3391/ai.2011.6.S1.024

7. Pereira D, Pinto M, Almeida JR, Correia-da-Silva M, Cidade H (2024) O Papel dos Flavonóides Naturais e Sintéticos na Prevenção da Bioincrustação Marinha. Mar Drugs 22(2):77. https://doi.org/10.3390/md22020077
8. Prousis KC, Kikionis S, Ioannou E, Morgana S, Faimali M, Piazza V et al (2022) Avaliação da Síntese e da Atividade Antiincrustante de Análogos de Bromosferol, um Diterpeno Bromado Isolado da Alga Vermelha Sphaerococcus coronopifolius. Mar Drugs 20(1):7. https://doi.org/10.3390/md20010007
9. Araújo CF, Batista AC, Gama BAP, Assis PS (2019) A Efetividade do Anti-incrustante de Algas Vermelhas para Reduzir o Impacto do Limnoperna fortunei – Mexilhão-dourado. In: Anais do 74° Congresso Anual da ABM, São Paulo, pp 2520–2529
10. Kumar GS, Shaik I, Singh HR, Sunil Kumar D, Reddy KVS, Dumpala S, Teja G (2023) Catequina como Agente Antiincrustante Natural contra Bactérias Formadoras de Biofilme da Costa de Visakhapatnam, Andhra Pradesh, Índia. Uttar Pradesh J Zool 44(23):1–10. https://doi.org/10.56557/upjoz/2023/v44i233755

Additive Manufacturing of TPU for Sports Eyewear Development and Characterization

Rogerio Rabello da Rocha Junior, Victor Paes Dias Gonçalves, Noam Tonini Simonassi, Sérgio Neves, Carlos Maurício Fontes Vieira, and Felipe Lopes Perisse

Abstract This work explores the application of additive manufacturing in the production of sports eyewear using thermoplastic polyurethane (TPU), a material known for its flexibility and resilience. 3D printing, such as fused deposition modeling (FDM), enables the creation of complex and customized geometries, ideal for eyewear designs that adapt to athletes' needs. The study focused on characterizing TPU to ensure the functionality and durability of the eyewear under various conditions of use. High-temperature dimensional stability tests were conducted, crucial to ensure the material maintains its shape and fit even when exposed to heat. Furthermore, sweat resistance was evaluated, aiming for product durability in contact with perspiration, an important factor for sports eyewear. The obtained results contribute to the advancement of additive manufacturing in the field of sports equipment, offering innovative and high-performance solutions.

Keywords Additive manufacturing · 3D printing · Thermoplastic polyurethane (TPU) · Sports eyewear

Introduction

The development of advanced materials has driven significant innovations in various industries, and sports equipment manufacturers have sought to embrace these innovations. The search for materials that offer an ideal combination of durability, flexibility, lightness, and safety has led to the growing use of engineering polymers. Among these, thermoplastic polyurethane (TPU) has stood out as a promising choice due

---

R. R. da Rocha Junior (✉) · V. P. D. Gonçalves · N. T. Simonassi · C. M. F. Vieira · F. L. Perisse
Advanced Materials Laboratory—LAMAV, State University of the Northern Rio de Janeiro—UENF, Campos dos Goytacazes, RJ, Brazil
e-mail: rogeriorabello30@gmail.com

S. Neves
Military Institute of Engineering (IME), Rio de Janeiro, Brazil

H. A. C. Lopera et al. (eds.), *Natural Fibers and Biocomposites 2026*, The Minerals, Metals & Materials Series, https://doi.org/10.1007/978-3-032-13577-3_13

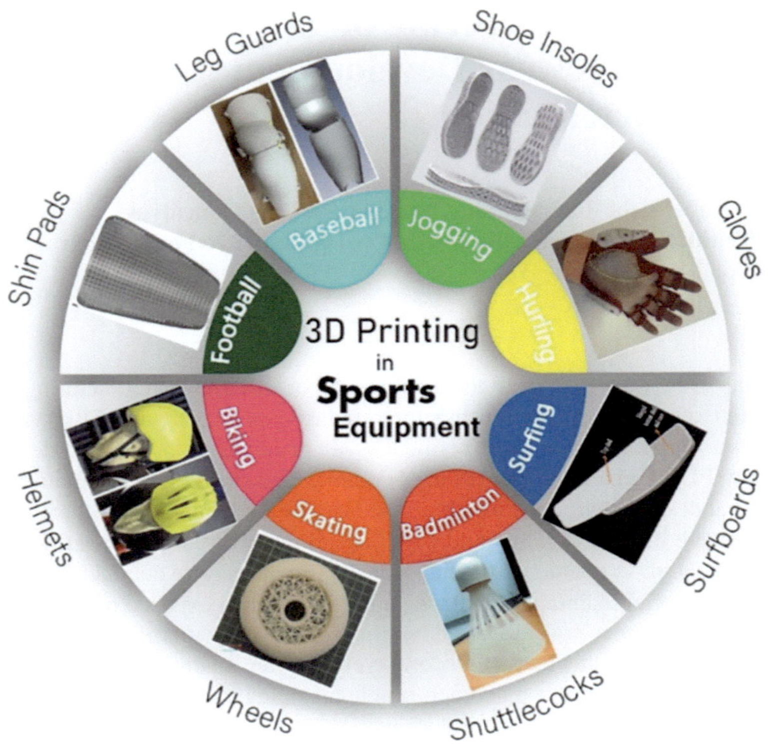

Fig. 1 Applications of 3D printing in sports equipment [1]

to its intrinsic properties that make it particularly suitable for demanding applications, such as sportsglasses, safety glasses, and everyday glasses that require superior resistance (Fig. 1).

High-performance eyewear is designed to protect the eyes from impact, UV radiation, dust, and other environmental elements, while providing comfort and visual clarity. The choice of material for the frame and, in some cases, for the lenses, is crucial to meeting these requirements. Traditional materials such as polycarbonate and acetate, although widely used, can have limitations in terms of flexibility, fatigue resistance, and impact absorption capacity in extreme conditions. TPU, with its unique polymer structure, offers an alternative that can overcome these limitations.

Thermoplastic polyurethane (TPU) is a thermoplastic elastomer that stands out for its unique combination of properties, including high elasticity, abrasion resistance, resistance to oils and greases, good flexibility at low temperatures, and excellent impact absorption capacity [2]. Chemically, TPU is a block copolymer consisting of hard segments and soft segments. The hard segments are typically formed by the reaction of a diisocyanate with a low molecular weight chain extender, while the soft segments are derived from a high molecular weight polyol. The ratio and nature of these segments determine the physical and mechanical properties of TPU, allowing for a wide range of hardness and flexibility [3].

The versatility of TPU is attributed to its biphasic structure, where rigid segments form crystalline or pseudo-crystalline domains that act as physical crosslinking points, providing strength and rigidity, while flexible segments provide elasticity and resilience. This microstructure allows TPU to be processed by conventional thermoplastic methods, such as injection molding, extrusion, and blow molding, making it a highly adaptable material for a variety of applications [4].

The demand for eyewear that offers superior protection, comfort, and durability has driven the adoption of TPU in the optical industry. The inherent properties of TPU make it an ideal material for sports eyewear frames, safety glasses, sunglasses, and even lens components. Its high impact resistance is crucial for safety and sports eyewear, where protection against shattering and collisions is paramount. TPU can effectively absorb and dissipate impact energy, minimizing the risk of eye injury [5].

In addition to impact resistance, the flexibility of TPU is a key factor in the comfort and fit of eyewear. Frames made of TPU can better adapt to the contours of the face, providing a secure and comfortable fit, even during intense physical activities. This flexibility also contributes to the durability of the product, as frames are less likely to break under mechanical stress or deformation [6].

Abrasion resistance and UV resistance are other important properties of TPU for eyewear applications. Continuous exposure to sunlight and harsh environments can degrade plastic materials, leading to cracking, discoloration, and loss of mechanical properties. TPU, especially UV-stabilized formulations, maintains its structural and aesthetic integrity for long periods, ensuring the longevity of eyewear [7]. The optical clarity of TPU is also an advantage, allowing the manufacture of transparent or translucent components that do not compromise the user's vision [8].

Examples of applications include ski goggles, swimming goggles, cycling goggles, and industrial safety goggles, where the combination of strength, flexibility, and durability is essential. TPU's ability to be molded into complex, ergonomic designs also allows for the creation of goggles with improved ventilation, fit, and aesthetics [9].

TPU can be processed using a variety of polymer manufacturing techniques, the most common being injection molding and extrusion. The choice of processing method depends on the component design, production volume, and desired properties of the final product. Injection molding is the most common method for manufacturing eyeglass frames and other complex components, allowing for the production of parts with high dimensional accuracy and quality surface finish [10]. Extrusion is used to produce continuous profiles, suitable for eyeglass temples [11]. More recently, 3D printing (additive manufacturing) has gained prominence for prototyping and customization [12]. Regardless of the method, proper drying of the material is crucial due to its hygroscopic nature [13].

This article aims to explore the potential of TPU in the manufacture of high-performance eyewear. Tensile testing will be detailed, which are essential for understanding the mechanical and thermal behavior of the material, respectively. A thorough understanding of these characteristics is vital for optimizing the design and functionality of eyewear that meets the growing demands for safety and performance.

Materials and Methods

For the study in question, white TPU Flex filament (Shore 95A) was used, supplied by 3D Fila [14]. The parts were printed with a direct-drive 3D printer, selected for its superior ability to handle flexible materials without clogging or kinking on the way to the hotend.

The experimental development adopted the following ideal printing parameters: nozzle temperature set at 220 °C; heated bed temperature set at 100 °C; printing speed of 60 mm/s (Fig. 2).

Samples for tensile testing were produced by additive manufacturing using White Flex TPU filament (Shore 95A), according to the manufacturer's specifications [14]. For this test, five test specimens of each composition were 3D printed with a thickness of 2 mm, following the geometry recommended by ASTM D638 [15].

The tensile test was conducted on a universal testing machine, comparing two different speed parameters: 50 and 500 mm/min, in order to evaluate the influence of the deformation rate on the mechanical properties of the material. For the Izod impact test, three specimens of each composition were printed with standardized dimensions according to ASTM D256 [16] and tested on an INSTRON. The fractured surfaces from the tensile test were analyzed by scanning electron microscopy (SEM).

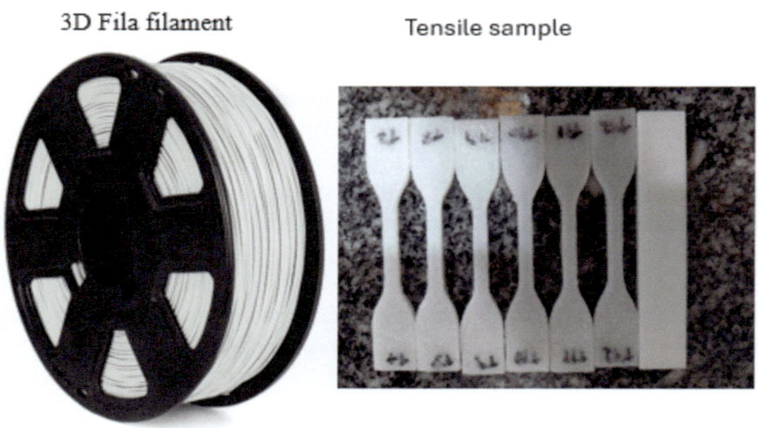

Fig. 2 Filament and tensile test specimens

Results and Discussion

The results of the tensile tests (Figs. 3 and 4) with TPU demonstrate that tensile strength is influenced by the test speed, with higher values at 50 mm/min compared to 500 mm/min. This behavior is related to the viscoelastic nature of the material, since at lower speeds the polymer chains have more time to reorganize and align in the direction of the applied stress, resulting in greater mechanical strength. On the other hand, at high deformation speeds, molecular mobility is limited, restricting the dissipation of internal stresses and leading to a reduction in strength. It can be observed that, in the first set of results, the strength reached about 22 MPa at 50 mm/min against approximately 16 MPa at 500 mm/min, while in the second set the values were higher, reaching 45 MPa at the lowest speed and falling to 30 MPa at the highest, with greater variation between the tests. Despite the absolute differences in resistance, possibly associated with different formulations or types of TPU, the trend observed was the same in both cases: higher mechanical performance at low speeds and reduced resistance as the deformation rate increases, confirming the typical behavior of viscoelastic polymers [17–19].

In Fig. 5, it can be observed that the material presented significantly different values between impact resistance and notch resistance. While impact resistance was relatively low, around 30 J/m^2, notch resistance reached very high values, close to 370 J/m^2. This difference demonstrates that the material has a high capacity to absorb energy when subjected to concentrated stresses in the notch region, highlighting its toughness and the influence of energy dissipation mechanisms during crack propagation [20].

Microscopic analysis (Fig. 6) of the fractured surface confirms this behavior, revealing irregular aspects and regions with elongation, typical of a fracture involving plastic deformation and energy absorption, which corroborates the results obtained

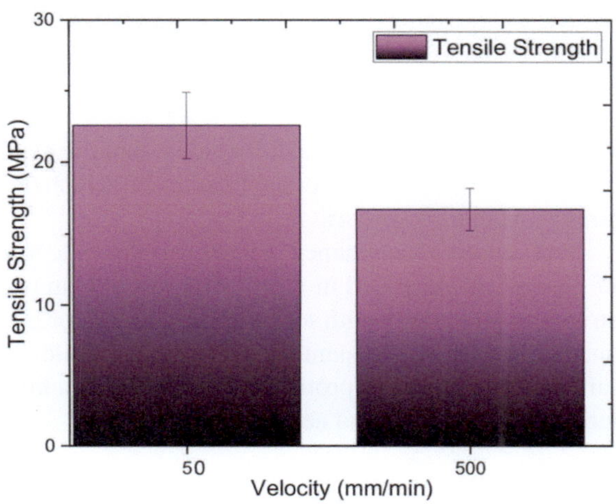

Fig. 3 Tensile strength

Fig. 4 Young's modulus

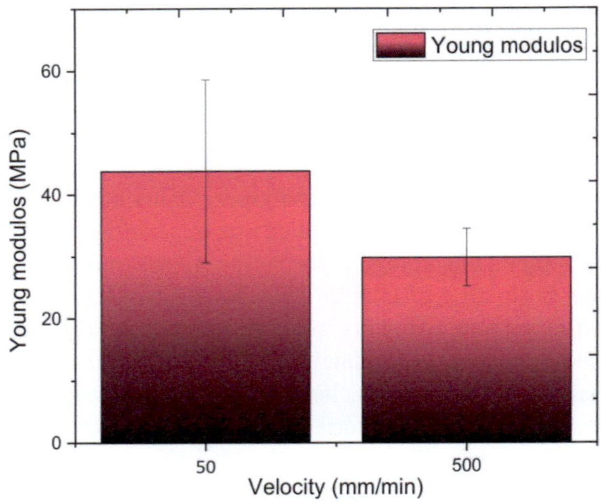

Fig. 5 Izod test

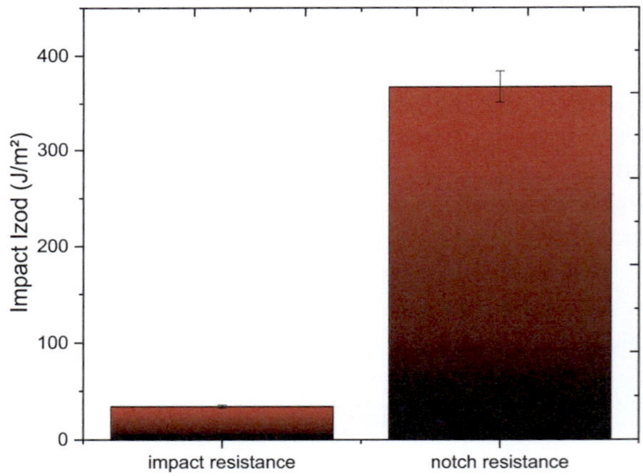

in the mechanical test. Thus, it is evident that the material has high notch resistance, a desirable characteristic for applications in which the component may be subject to stress concentrations during use [18, 21].

In addition to the planned tests, it was possible to develop a functional prototype of glasses, as illustrated in Fig. 7. The model features a modern design, lightweight structure, and lenses with technology suitable for eye protection, ensuring comfort and performance. The manufacturing process followed the steps of modeling, structural optimization, and prototyping, resulting in a final product that is satisfactory in terms of ergonomics and aesthetics.

Fig. 6 Impact resistance and microscopy of the fractured sample

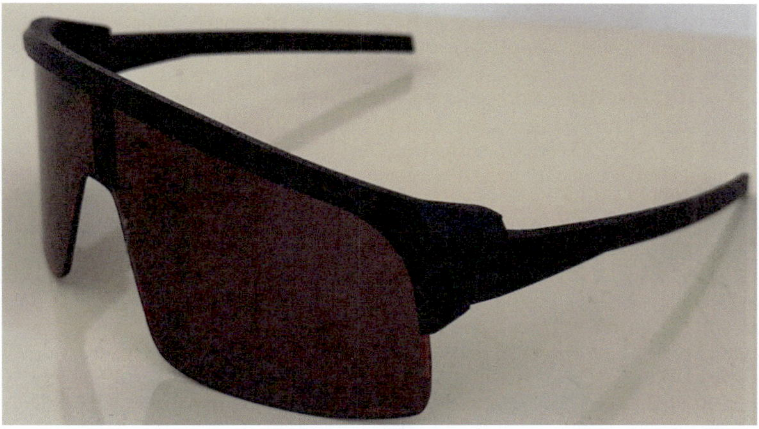

Fig. 7 Prototype glasses developed, featuring a lightweight frame and lenses with optical filtering technology

Conclusion

This study demonstrated the potential of using thermoplastic polyurethane (TPU) combined with additive manufacturing in the production of high-performance sports glasses. Mechanical tests confirmed the viscoelastic nature of the material, showing greater resistance at lower deformation rates and highlighting its high toughness in the face of stress concentrations. In addition, thermal stability and sweat resistance tests reinforce the suitability of TPU for applications in real-world conditions.

Based on the characterization performed, it was possible to develop a functional prototype of glasses, combining lightness, ergonomic comfort, and aesthetic performance, highlighting the feasibility of the modeling, structural optimization, and prototyping process via 3D printing. Thus, this work contributes to the advancement of additive manufacturing applications in the sports equipment sector, offering an innovative and customized alternative to conventional materials and methods.

Acknowledgements CNPQ (proq. n. 301634/2018-1) for the financial support and FAPERJ through the projects Ref. Proc. n. E-26/203.874/2022.

References

1. Namvar N, Moloukzadeh I, Zolfagharian A, Demoly F, Bodaghi M et al (2023) Bio-inspired design, modeling, and 3D printing of lattice-based scale model scooter decks. Int J Adv Manuf Technol 15(126):2887–2903
2. Saeedi S, Ghasemi I (2020) Investigating physio-thermo-mechanical properties of polyurethane elastomers. J Polym Res 27(8):241
3. Li J et al (2025) Self-reinforced thermoplastic polyurethane composite with excellent mechanical properties, thermal stability and sustainable recycling. Compos Part B Eng 295:108090
4. Liu H et al (2025) Preparation and application of high strength, high transparency, and high impact resistance polyurethane elastomers by controlled cross-linking. Polymer 315:123387
5. Huang G et al (2023) Colorless, transparent, and high-performance polyurethane with intrinsic ultraviolet resistance and its anti-UV mechanism. ACS Appl Mater Interfaces 15(11):14207–14216
6. Bozukova D et al (2015) Evaluation of a class of polyurethane materials for intraocular lens manufacturing. J Biomed Mater Res B Appl Biomater 103(2):380–388
7. Xu T, Shen W, Lin X, Xie YM (2020) Mechanical properties of additively manufactured thermoplastic polyurethane (TPU) material affected by various processing parameters. Polymers (Basel) 12(12):3010
8. Al-Mulla A, Al-Marzouqi AH (2024) Graphene nanocomposite for 3D printed eyewear. Khalifa University
9. Duncan O et al (2018) Review of auxetic materials for sports applications: expanding options in comfort and protection. Appl Sci 8(6):941
10. Yan G et al (2025) From molecular design to scenario adaptation: cutting-edge exploration of silicone-modified polyurethane in smart sports fields. Coatings 15(7):737
11. Kim D, Choi Y (2021) Applications of smart glasses in applied sciences: a systematic review. Appl Sci 11(11):4956
12. Sun X et al (2024) Flexible thermoplastic polyurethane composites with ultraviolet resistance for fused deposition modeling 3D printing. 3D Print Addit Manuf
13. Saputra NI, Ramadhani HK (2024) Safety goggles myopi as a worker protector with myopic eye complaints. Malays J Public Health Commun Med 2(1):5859
14. 3D Fila. Filamento TPU Flex Branco [Internet]. 3D Fila, São Paulo. https://3dfila.com.br/produto/filamento-tpu-flex-branco/. Accessed 3 Sept 2025
15. ASTM International (2014) ASTM D638-14: standard test method for tensile properties of plastics. ASTM International, West Conshohocken, PA

16. ASTM International (2024) ASTM D256-24: standard test methods for determining the izod pendulum impact resistance of plastics. ASTM International, West Conshohocken, PA
17. Chaudhry MS, Czekanski A (2020) Evaluating FDM process parameter sensitive mechanical performance of elastomers at various strain rates of loading. Materials (Basel) 13(14):3202
18. Vidakis N, Petousis M, Korlos A, Velidakis E, Mountakis N, Charou C, Myftari A (2021) Strain rate sensitivity of polycarbonate and thermoplastic polyurethane for various 3D printing temperatures and layer heights. Polymers (Basel) 13(16):2752. https://doi.org/10.3390/polym13162752
19. Miao Y-G, He H, Li Z (2020) Strain hardening behaviors and mechanisms of polyurethane under various strain rate loading. Polym Eng Sci 60(5):1083–1092. https://doi.org/10.1002/pen.25364
20. Chaudhry MS, Czekanski A (2020) Evaluating FDM process parameter sensitive mechanical performance of elastomers at various strain rates of loading. Materials (Basel) 13(14):3202. https://doi.org/10.3390/ma13143202
21. Miao YG, He H, Li Z (2020) Strain hardening behaviors and mechanisms of polyurethane under various strain rate loading. Polym Eng Sci 60(5):1083–1092. https://doi.org/10.1002/pen.25364

Characterization and Reuse Potential of High-Density Polyethylene (HDPE) from Decommissioned Oil and Gas Industry Pipelines

Pedro Henrique Rabello Neves, Victor Paes Dias Gonçalves, Arthur Camposo Pereira, Noan Tonini Simonassi, Carlos Maurício Fontes Vieira, and Felipe Perisse Duarte Lopes

Abstract The inadequate disposal of polymeric materials, such as High-Density Polyethylene (HDPE) present in oil and gas industry pipelines, represents a significant environmental challenge. The search for effective and sustainable reuse solutions is, therefore, essential. This research evaluates the mechanical, physical, and chemical properties of recovered High-Density Polyethylene, originating from the same location and decommissioning context as the original material. For this purpose, comprehensive tests will be conducted: Shore D hardness, dilatometry, differential scanning calorimetry (DSC), and thermogravimetric analysis (TGA). The results indicated that the post-use High-Density Polyethylene maintains adequate properties, validating its potential for successful reintroduction into the production chain. This approach aims not only to mitigate the environmental impact of waste disposal but also to generate substantial economic value, transforming a liability into a valuable resource. The material's viability and the promotion of the circular economy are central pillars of this study.

Keywords Decommissioning · Polymer · Sustainability · High-Density Polyethylene (HDPE) · Reuse

P. H. R. Neves (✉) · V. P. D. Gonçalves · N. T. Simonassi · C. M. F. Vieira · F. P. D. Lopes
Advanced Materials Laboratory—LAMAV, State University of the Northern Rio de Janeiro—UENF, Campos dos Goytacazes, RJ, Brazil
e-mail: 20201100088@pq.uenf.br

A. C. Pereira
Metallurgical Engineering Federal University of Ouro Preto, Ouro Preto, MG, Brazil

H. A. C. Lopera et al. (eds.), *Natural Fibers and Biocomposites 2026*, The Minerals, Metals & Materials Series, https://doi.org/10.1007/978-3-032-13577-3_14

Introduction

The accelerated growth of offshore decommissioning activities represents one of the major environmental and technological challenges of the present time [1, 2]. In the coming decades, thousands of kilometers of pipelines and associated components are expected to be removed from service, resulting in a significant volume of polymeric and metallic waste. Among these polymers, High-Density Polyethylene (HDPE) is widely used as an external protective layer in flexible pipelines, providing durability and resistance to marine environments [1]. At the end of its service life, however, this material, due to its high chemical stability, becomes a difficult-to-manage environmental liability, persisting for long periods in the environment and contributing to the generation of microplastics [3].

The proper management of these wastes is not only an environmental requirement but also a strategic demand for the oil and gas industry, which faces increasing regulatory and societal pressure to adopt practices aligned with sustainability. Conventional routes such as landfilling or incineration, besides being limited, are associated with high costs and significant greenhouse gas emissions. The production of virgin petrochemical HDPE itself is highly impactful: it is estimated that each ton of HDPE produced may emit between 1.8 and 3.2 tons of CO_2 equivalent, contributing substantially to global emissions, which reached about 1.9 ± 0.6 Gt CO_2 in 2020 [4]. In this context, the reuse of post-service HDPE emerges as a promising alternative, capable of reducing environmental impacts while generating economic value from a high-volume waste stream [2, 5].

Embedded within the principles of the circular economy, the recycling of HDPE from decommissioned pipelines enables its reintegration into production chains, partially replacing virgin resins [5] and contributing to the mitigation of CO_2 emissions associated with petrochemical production [2, 5]. Furthermore, this approach strengthens the industry's sustainable image, reduces liabilities, and maximizes the use of existing resources [2, 5].

Although recycled HDPE can be directed to various industrial sectors, such as technical packaging, fiber-reinforced composites, and other low-demand structural components [6], this study focuses on its potential use in additive manufacturing processes, particularly in the production of filaments for fused deposition modeling (FDM). Recent research has demonstrated that recycled HDPE can be transformed into filaments for 3D printing, showing adequate mechanical performance and even improvements after multiple recycling cycles before degradation begins [7]. This direction not only highlights the feasibility of HDPE reuse but also its potential insertion into an expanding technological context, combining design flexibility, manufacturing innovation, and sustainability.

Methodology

Material and Preparation

The polymeric material investigated in this study was High-Density Polyethylene (HDPE) collected from the *outer sheath* layer of flexible subsea pipelines subjected to decommissioning (Fig. 1). The material was obtained as raw segments of the external protective coating. Although the precise origin of the pipelines could not be traced, the average service lifetime was estimated to be approximately 20 years. The samples were sourced from the Porto do Açu region (RJ, Brazil), an area of intense offshore activity that generates significant volumes of end-of-life infrastructure materials.

Two material conditions were selected for comparison: as-recovered (in natura), which refers to samples taken directly after decommissioning without additional thermal or mechanical treatment, and reprocessed HDPE, which underwent a thermo-mechanical cycle simulating a conventional recycling process. For the reprocessed condition, the raw HDPE was mechanically fragmented in a knife mill, and the resulting particles were consolidated in a metallic mold after thermal treatment at 180 °C for 1 h, yielding unified specimens (Fig. 2). This procedure was designed to replicate a typical thermal reprocessing step.

Fig. 1 HDPE segments recovered from decommissioned pipelines

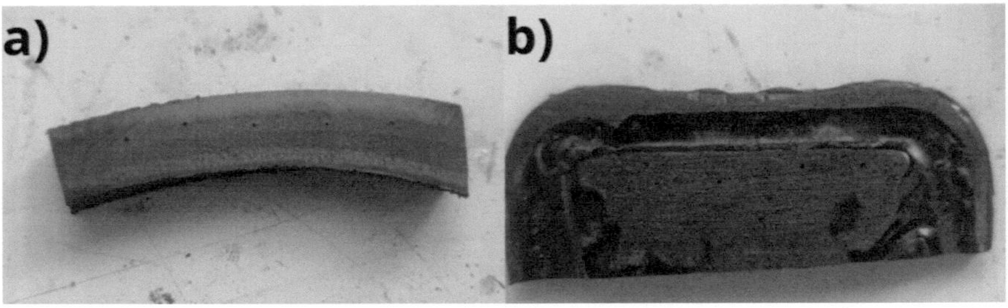

Fig. 2 Specimens for Shore D hardness: **a** as-recovered and **b** reprocessed

Characterization Techniques

Shore D Hardness

Hardness testing was performed according to ASTM D2240 [8]. Five independent measurements were conducted on each specimen using a Shore D durometer, and the average values were reported as representative for each condition.

Dilatometry

The coefficient of linear thermal expansion (CTE) was measured using a NETZSCH DIL 402 PC dilatometer (LAMAV/UENF), following ASTM E228 [9]. Tests were conducted on the as-recovered HDPE specimens, which best represent the original in-service condition of the material. Samples were subjected to a heating program from 5 to 150 °C at a rate of 2 °C/min.

Thermogravimetric Analysis (TGA) and Differential Scanning Calorimetry (DSC)

Thermal stability and melting behavior were assessed exclusively on the as-recovered HDPE specimens, to characterize the material's pristine thermal response prior to reprocessing. TGA was carried out in accordance with ASTM E1131 [10], using a TA Instruments system (New Castle, DE, USA). Samples were stabilized at 30 °C for 5 min, then heated up to 600 °C at a rate of 10 °C/min under nitrogen (N_2) atmosphere. Prior to testing, all samples were stored in a desiccated environment. The thermograms were processed using TA Universal Analysis software.

Results and Discussion

The Shore D hardness analysis (Fig. 4) allowed the evaluation of the mechanical behavior of HDPE after long-term exposure in the offshore environment and following thermal reprocessing. The hardness values of the *as-recovered* and reprocessed conditions were very similar, with only small variations within the standard deviation. This stability indicates that, even after approximately 20 years of service and a subsequent reprocessing step, the polymer chains remained structurally stable, without significant compromise of surface rigidity (Fig. 3).

The dilatometric analysis (Fig. 5) revealed a typical thermal expansion profile for HDPE. A linear expansion behavior was observed with increasing temperature up to approximately 100 °C, consistent with the viscoelastic regime of the polymer. Above this range, a more pronounced dilation occurred, reaching a peak near 120 °C, close to the melting transition of HDPE.

The thermogravimetric curve (TGA) (Fig. 5) demonstrated high thermal stability of the material up to approximately 400 °C, with a sharp mass loss beyond this point, indicating the onset of HDPE thermal decomposition. This result agrees with the degradation behavior reported in the literature, which is characterized by a single main decomposition event and suggests the absence of significant organic contaminants or mineral fillers from offshore use [11]. The DSC analysis (Fig. 5) corroborates these findings, revealing an endothermic melting peak near 130 °C, consistent with the behavior expected for highly crystalline HDPE [12]. These results are also in line with the Shore D hardness test, where a slight reduction in the hardness of reprocessed samples was attributed to the onset of thermal degradation during heating at 180 °C.

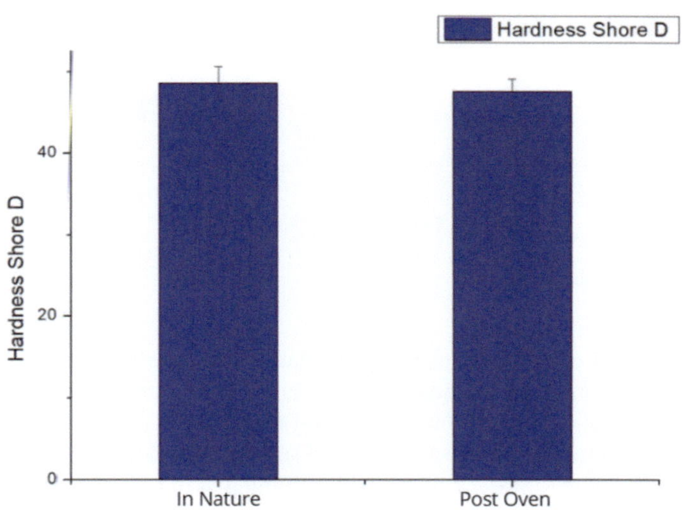

Fig. 3 Shore D hardness

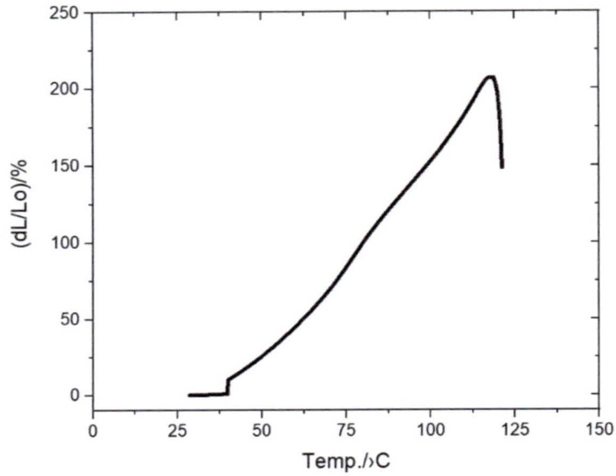

Fig. 4 Dilatometry results

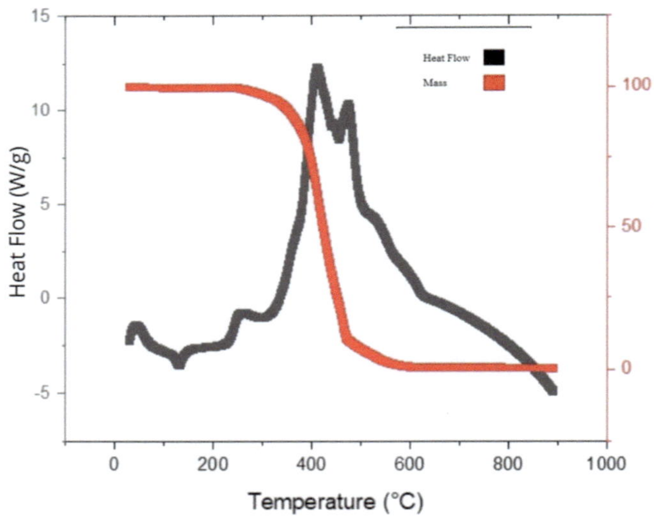

Fig. 5 TGA–DSC results

The integration of the findings confirms that HDPE recovered from decommissioned pipelines maintains adequate thermal and mechanical properties for reuse, particularly in applications of lower structural demand. Its thermal stability, resistance to degradation, and dimensional control support the feasibility of direct reprocessing, adding value to waste streams while aligning with the principles of sustainability and circular economy in the offshore sector. Among the feasible applications, the conversion of HDPE into filaments for additive manufacturing via fused deposition modeling (FDM) stands out, enabling the recycled polymer to be used in the fabrication of technical parts, prototypes, and industrial components [13, 14].

Discussion

This study demonstrated the feasibility of reusing High-Density Polyethylene (HDPE) recovered from decommissioned subsea pipelines in the offshore sector. Even after nearly two decades of exposure to aggressive marine and operational conditions, the material preserved its mechanical and thermal integrity, as confirmed by hardness, dilatometry, TGA, and DSC analyses. The results revealed stable performance of the polymer chains, with only minor variations after reprocessing, indicating that the material remains structurally reliable.

Beyond hardness and thermal stability, additional characterizations performed in this work such as tensile strength, density, FTIR, and wettability provided a broader understanding of the physicochemical behavior of the recycled polymer. These analyses confirmed that no severe oxidation or contaminant uptake occurred during service life, and that the surface and bulk properties remain adequate for reuse.

The integration of these findings highlights that direct reprocessing of HDPE from decommissioned pipelines represents a viable and sustainable route, enabling the valorization of high-volume polymeric residues from the oil and gas industry. In addition to reducing environmental impacts associated with improper disposal and virgin resin production, this approach aligns with the principles of the circular economy, offering a pathway toward resource efficiency and emissions reduction.

Among the most promising applications, the conversion of recycled HDPE into filaments for additive manufacturing via fused deposition modeling (FDM) stands out. The preservation of dimensional stability, thermal resistance, and mechanical strength supports the suitability of this polymer for the fabrication of technical components, prototypes, and other products with lower structural demand. Therefore, this study provides strong evidence that decommissioned HDPE can be successfully reintegrated into new production cycles, coupling technological innovation with sustainability in the offshore industry.

Conclusion

This study demonstrated the potential for reusing High-Density Polyethylene (HDPE) from decommissioned subsea pipelines in the offshore sector. The analyses showed that, even after approximately two decades of exposure to severe operating conditions, the material maintains Shore hardness properties after reprocessing.

Acknowledgements The authors thank the Brazilian agencies CNPq, process number 302976/2022-1, and FAPERJ, process number E-26/200.847/2021.

References

1. Oluwoye I, Machuca LL, Higgins S, Suh S, Galloway TS, Halley P, Tanaka S, Iannuzzi M (2023) Degradation and lifetime prediction of plastics in subsea and offshore infrastructures. Sci Total Environ 904:166719. https://doi.org/10.1016/j.scitotenv.2023.166719
2. Capobianco N, Basile V, Loia F, Vona R (2022) End-of-life management of oil and gas offshore platforms: challenges and opportunities for sustainable decommissioning. Sinergie Ital J Manag 40(2):299–326. https://doi.org/10.7433/s118.2022.14
3. Duong TT, Le PT, Nguyen TNH, Hoang TQ, Ngo HM, Doan TO, Le TPQ, Bui HT, Bui MH, Trinh VT, Nguyen TL, Vu TM, Tran TKC, Ho TC, Phuong NN, Strady E (2022) Selection of a density separation solution to study microplastics in tropical riverine sediment. Environ Monit Assess 194:65. https://doi.org/10.1007/s10661-021-09664-0
4. Cullen L, Meng F, Lupton R, Cullen JM (2024) Reducing uncertainties in greenhouse gas emissions from chemical production. Nat Chem Eng. https://doi.org/10.1038/s44286-024-00047-z
5. Beena Unni A, Joseph TM (2024) Enhancing polymer sustainability: eco-conscious strategies. Polymers 16:1769. https://doi.org/10.3390/polym16131769
6. El Hawary O, Boccarusso L, Durante M, Panico M, Pinto F (2025) Production, repairing and testing of hemp reinforced recycled HDPE composites. Arab J Sci Eng. https://doi.org/10.1007/s13369-025-10170-5
7. Khan JA, Noon AA, Arif M, Naveed M, Khan HA, Suhaib MM, Sharif A (2023) Mechanical properties evaluation of recycled high-density polyethylene via additive manufacturing. J Mater Manuf 2:1–9. https://doi.org/10.5281/zenodo.10012303
8. ASTM D2240-15 (2015) Standard test method for rubber property—durometer hardness. ASTM International, West Conshohocken, PA
9. ASTM E228 (2022) Standard test method for linear thermal expansion of solid materials with a push-rod dilatometer. ASTM International, West Conshohocken, PA
10. ASTM International (2023) ASTM D3878-23: standard terminology for composite materials. ASTM International, West Conshohocken, PA
11. Costa CS, Fernandes A, Munoz M, Ribeiro MR, Silva JM (2024) Analyzing HDPE thermal and catalytic degradation in hydrogen atmosphere: a model-free approach to the activation energy. Catalysts 14:514. https://doi.org/10.3390/catal14080514
12. De Monte C, Locritani M, Merlino S, Ricci L, Pistolesi A, Bronco S (2022) An in situ experiment to evaluate the aging and degradation phenomena induced by marine environment conditions on commercial plastic granules. Polymers 14:1111. https://doi.org/10.3390/polym14061111
13. Handwerker M, Wellnitz J, Marzbani H, Tetzlaff U (2021) Annealing of polyamide 6 reinforced with chopped and continuous fibers produced by fused filament fabrication. Compos Part B Eng 223:109119. https://doi.org/10.1016/j.compositesb.2021.109119
14. Olawumi MA, Oladapo BI, Olugbade TO (2024) Evaluating the impact of recycling on polymer of 3D printing for energy and material sustainability. Resour Conserv Recycl 209:107769. https://doi.org/10.1016/j.resconrec.2024.107769

Characterization of Lignocellulosic Fibers from Açaí Palm Stem (*Euterpe oleracea*) for Potential Use in Polymer Composites

Marcela Coelho da Silva, Leonardo Martins Nazário Silva, Heliane Rosa do Amaral, Felipe Gabriel Santos Araújo, Hemanuelly Ferreira Breda Lan Oliveira, Felipe Perissé Duarte Lopes, and Michel Picanço Oliveira

Abstract This study explores the potential of açaí palm stem fibers (AFPS) as reinforcement in composite materials, aligning with circular economy principles by converting agricultural waste into a valuable resource. AFPS present a sustainable alternative to synthetic fibers such as fiberglass and nylon, owing to their low density, high specific strength and stiffness, biodegradability, and carbon neutrality. As Brazil—the world's largest açaí producer—faces increasing waste from fruit processing, the valorization of discarded stems offers an environmentally responsible solution. In this research, AFPS extracted from waste stems were characterized in terms of their mechanical (tensile strength), thermal (TGA), microstructural (FTIR), morphological (SEM), chemical, and physical properties. The fibers exhibited a maximum tensile strength of 267.3 MPa and an initial thermal degradation temperature of 278 °C. These results demonstrate the feasibility of using AFPS in composite applications, reducing dependence on non-renewable resources while promoting sustainable waste management.

M. C. da Silva · L. M. N. Silva · H. R. do Amaral · F. G. S. Araújo · H. F. B. L. Oliveira · M. P. Oliveira (✉)
Forest and Wood Sciences Department, Federal University of Espírito Santo, Jeronimo Monteiro, Brazil
e-mail: michel.oliveira@ufes.br

M. C. da Silva
e-mail: marcela.c.silva@edu.ufes.br

L. M. N. Silva
e-mail: leonardo.mn.silva@edu.ufes.br

F. G. S. Araújo
e-mail: felipe.g.araujo@edu.ufes.br

F. P. D. Lopes
Materials Science and Engineering Department, State University of Northern Rio de Janeiro, Campos dos Goytacazes, Brazil
e-mail: perisse@uenf.br

H. A. C. Lopera et al. (eds.), *Natural Fibers and Biocomposites 2026*, The Minerals, Metals & Materials Series, https://doi.org/10.1007/978-3-032-13577-3_15

Keywords Physicochemical characterization · *Euterpe oleracea* · Composite

Introduction

The açaí palm tree (*Euterpe oleracea*), native to the Amazon region, yields a dark purple fruit highly valued in northern Brazil. Açaí is one of the Amazonian products that has experienced significant growth in both supply and, especially, demand, expanding into national and international markets over the past two decades.

The product's strong market acceptance is attributed to its nutritional and functional characteristics, as it contains considerable levels of anthocyanins and unsaturated fatty acids, among other distinctive attributes [1]. The pulp of the fruit is used to produce juice, which serves as both a food source and a component in dietary supplements and cosmetics. Açaí production, whether agricultural or extractive, is environmentally significant because, when conducted sustainably, it can aid in the conservation of the Amazon rainforest. From a social perspective, this activity is predominantly carried out by traditional communities and family farmers, often serving as one of their main sources of income. It is estimated that the extraction, transportation, commercialization, and industrialization of açaí fruits and palm hearts generate 25,000 direct jobs and contribute over R$ 40 million in annual revenue [2, 3].

The açaí palm stipe exhibits a range of characteristics that make it a promising candidate for use in natural fiber composites. Its smooth, cylindrical, and unbranched structure, growing up to 30 m tall with diameters of 12–18 cm, suggests the potential for long, strong fibers. The fibrous nature of the stipe, along with its sometimes curved shape, could offer unique advantages in composite applications where flexibility and structural integrity are desired [1–3].

However, several factors need to be explored further to assess the suitability of açaí palm fibers for composite materials. These include understanding the mechanical properties of the fibers, comparing them to commonly used natural fibers, and evaluating the scalability of fiber production. Additionally, the potential challenges posed by the curved shape of the stipe and the variability in fiber quality due to node and internode structures should be investigated.

Experimental

Materials

The palm trees were harvested in the interior of the southern region of Espírito Santo state, in the city of Rio Novo do Sul, by a local producer who does not utilize the stems after harvesting. The fibers were manually extracted and stored in a SolidSteel sterilization and drying oven to prevent the proliferation of fungi and mold.

Characterizations

After separating the fibers, they were analyzed using a confocal microscope (Laser Scanning Microscope for Material Analysis LEXT OLS5100). This analysis was conducted to measure the fiber diameters, with the objective of evaluating the diametric distribution. To construct the histogram, Eqs. 1, 2, and 3 were utilized. Equation 1 was used to determine the number of classes in the histogram, while Eq. 2 established the range of values. Finally, Eq. 3 allowed for the calculation of the cutoff values for each class.

$$C = \sqrt{N} \tag{1}$$

$$At = F_{\text{major}} - F_{\text{minor}} \tag{2}$$

$$Ac = \frac{At}{C - 1} \tag{3}$$

where C—Number of classes; N—Number of samples; At—Amplitude of the sample group; Ac—Cutoff value; F_{major} and F_{minor}—Fiber size in μm.

Scanning electron microscopy (SEM) images were acquired on a JEOL JSM-IT200 microscope. Samples were fixed, sputter-coated with a thin layer of gold and examined at an accelerating voltage of 10 kV.

Fourier-transform infrared (FTIR) spectra were recorded in attenuated total reflectance (ATR) mode on a Bruker Tensor 27 spectrometer over the 4000–600 cm^{-1} region with 4 cm^{-1} resolution.

Thermogravimetric analysis (TGA) was carried out on a Setaram LabSys Evo apparatus by heating specimens from 25 to 700 °C at 10 °C/min under a nitrogen atmosphere.

X-ray diffraction (XRD) patterns of FIN and FAC samples were obtained using a Rigaku MiniFlex 600 diffractometer with Cu Kα radiation ($\lambda = 1.5418$ Å), scanning 2θ from 3° to 70° at 0.05° steps and 2°/min. Powdered samples were also analyzed on a Shimadzu XRD-6000 under identical conditions to determine the crystallinity index (CI). CI was calculated according to: [4]

$$IC = \frac{I_{(002)} - I_{(\text{am})}}{I_{(002)}} * 100 \tag{4}$$

where I_{002} is the intensity of the (002) crystalline peak and I_{am} the amorphous background intensity.

Mechanical tensile tests were performed on an Oswaldo Filizola AME-2 kN tester with a gauge length (L_0) of 115 mm.

The chemical analyses of the fiber extracted from the açaí palm stem (AFPS) were performed according to standard TAPPI [5] methods in an integrated, sequential manner. First, moisture content was determined gravimetrically on conditioned, milled samples (40–60 mesh) by weighing, drying at 100 °C for 24 h, and reweighing. Next, holocellulose was isolated from extractive-free samples following Wise et al. [6]: the material was filtered, dried at 70 °C for 8 h, then weighed after cooling. A portion of this holocellulose was treated with 17.5% (w/v) KOH to yield α-cellulose, calculated from the difference between initial and final masses corrected for extractives and lignin. Hemicellulose content was obtained by subtracting α-cellulose from holocellulose. Insoluble lignin was quantified on about 0.3 g of extractive-free sample subjected to acid hydrolysis, filtered and dried at 100 °C, while the soluble lignin fraction was measured by UV–Vis spectrophotometry at 215 and 280 nm. Total lignin was reported as the sum of both fractions. Finally, extractives were removed in a Soxhlet apparatus with an ethanol–toluene mixture (1:2 v/v) for 5 h, followed by pure ethanol for 4 h, and their content determined by mass loss.

All results are expressed on a dry basis and according to absorbance readings, in accordance with TAPPI protocols. The results obtained were as follows: moisture content of 6.86%, lignin content of 14.77%, cellulose content of 63.37%, hemicellulose content of 29.56%, and holocellulose content of 92.93%.

Results and Discussion

Influence of Fiber Diameter on Mechanical Performance

Figure 1 presents a histogram of the measured cut-thickness values (in micrometers) for approximately 120 fibers. The distribution exhibits a bell-shaped profile, with the highest frequency concentrated in the 1100–1200 μm and a fitted mean of approximately 1200 μm. The narrow spread, representing about ±15% of the mean, demonstrates the reproducibility of the process. A single tail-end outlier near 1800 μm suggests an occasional anomaly in the cutting procedure, such as tool wear or feed-rate fluctuations. However, the data align well with the normal model, indicating that random variation predominates over systematic bias. In composite-based materials, where uniform thickness is crucial for ensuring consistent mechanical or thermal properties, this analysis confirms that our fabrication protocol produces highly homogeneous fibers materials. No fibers were obtained in the collection intervals corresponding to samples 8 and 9. Therefore, these samples were excluded from all subsequent analyses, and their results are not presented in any of the characterization sections.

The mechanical characterization (Fig. 2), of the ten fiber classes, classified through statistical analysis (Fig. 1), demonstrates a clear hierarchy of strength, stiffness, and toughness, which exhibits a direct correlation with fiber diameter. Classes 2–4, which exhibit the highest mean ultimate tensile strengths (318–331 MPa) and elastic moduli (19.6–20.1 GPa), correspond to the smallest average fiber diameters. Conversely, Classes 7 and 10, with the largest diameters, display the lowest strengths (180–159 MPa) and stiffness's (9.8–7.1 GPa). This inverse relationship between fiber

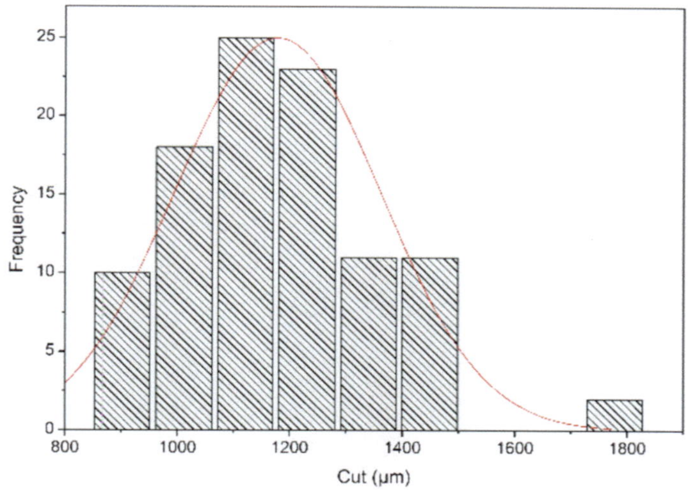

Fig. 1 Statistical distribution of section thickness (µm) with Gaussian fit

diameter and both tensile strength and modulus of elasticity is a well-documented phenomenon in lignocellulosic materials [7, 8].

These results suggest that reducing fiber diameter decreases the probability of critical defects and promotes more uniform molecular or crystalline alignment along the fiber axis, thereby enhancing load transfer and reducing local stress concentrations. In our data, Class 4 achieves the peak tenacity (2.98 MJ/m^3) despite a slightly lower modulus than Classes 2–3, suggesting an optimal balance of ductility and stiffness at an intermediate diameter. Classes with very fine diameters (2–3) maintain excellent stiffness but sacrifice some toughness, whereas overly large diameters (7–10) compromise all three mechanical metrics.

Physicochemical Characterization of AFPS

The SEM images of the AFPS surface are shown in Fig. 3.

First, it is interesting to note the distinct morphology of the AFPS surface, which is composed of small fragments of irregular shapes and with no evident indication of a fibrillar structure (Fig. 3d) in contrast with the results obtained for the Fig. 3c the image recorded with enhanced magnification indicates the presence of a number of features (such as terraces and steps) that are typical of the surface of lignocellulosic fibers [9]. On the other hand, Fig. 3a, b shows images recorded for AFPS surface in several areas. These images show that the core is composed by structures of different shapes and sizes. Figure 3b clearly evidence the presence of macrofibers

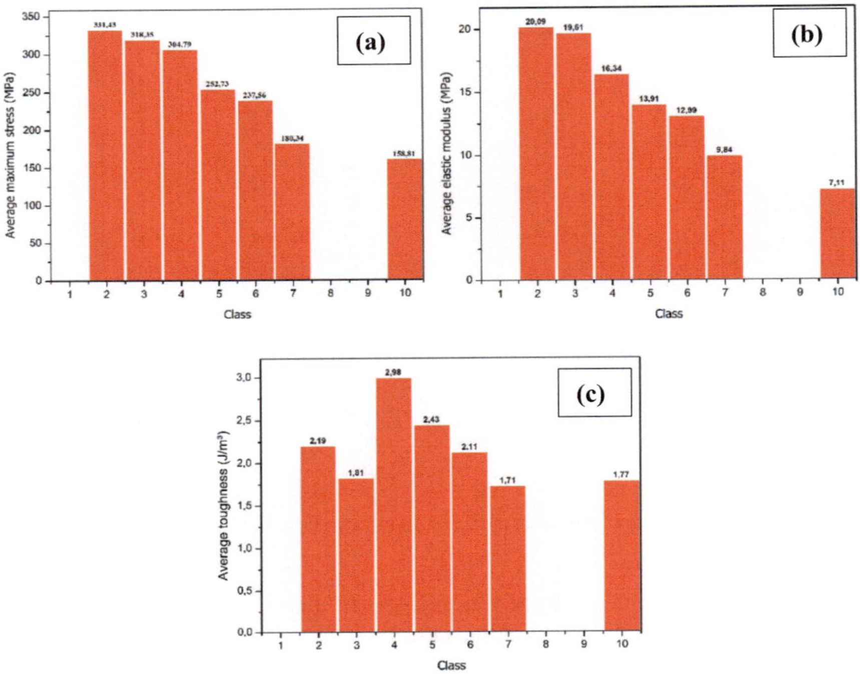

Fig. 2 **a** Maximum stress analysis performed across various material classes. **b** Elastic modulus distribution and mechanical performance variation observed. **c** Average toughness determined for each material classes

(with diameters ~100 μm or larger) formed by bundles of microfibers (with diameters ~10 μm).

Figure 4a shows the FTIR spectrum of the AFPS.

Based on the spectrum presented in Fig. 4a, a broad absorption band centered at 3312 cm^{-1} is observed, which corresponds to intermolecular O–H stretching vibrations. This band is typically associated with alcohols, extractives, and carboxylic acids that constitute cellulose and contribute to its hydrophilic character [10]. The band at 2889 cm^{-1} is attributed to C–H stretching vibrations from natural compounds present in cellulose and hemicellulose [8]. The absorption at 1720 cm^{-1} corresponds to the C=O stretching of carbonyl groups, while the band at 1597 cm^{-1} also represents C=O stretching vibrations [11–14]. The peak observed at 1422 cm^{-1} may be related to symmetric in-plane bending of CH_2 groups [10]. The band at 1366 cm^{-1} is assigned to C–H bending in cellulose [10, 15]. The occurrence of the band at 1236 cm^{-1} is associated with C–O–C stretching vibrations in aliphatic and aromatic groups [10], whereas the band at 1154 cm^{-1} corresponds to angular deformation of C–O bonds in esters present in the fibers. The prominent band at 1031 cm^{-1} corresponds to C–O stretching vibrations, characteristic of crystalline cellulose and often

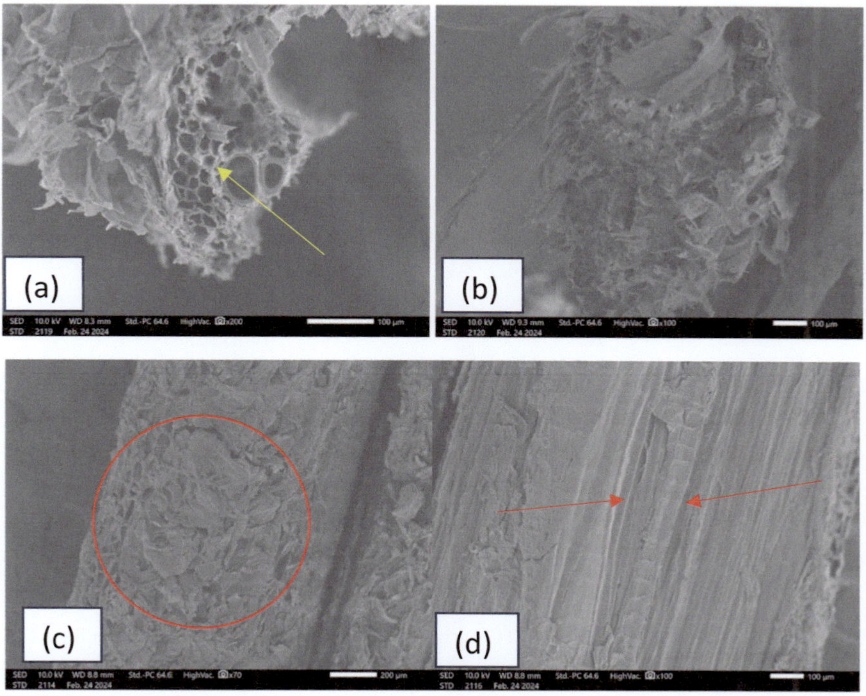

Fig. 3 SEM images of the AFPS: **a**, **b** core region and **c**, **d** surface area, shown at two different magnifications images

associated with its purity [16–18]. The band at 896 cm^{-1} indicates axial deformation of C–O–C bonds and β-glycosidic linkages between glucose units in cellulose [16–18]. Finally, the absorption at 699 cm^{-1} is attributed to out-of-plane bending of C–H bonds in aromatic rings [19, 20].

The thermal behavior of the AFPS was studied by TG, as shown in Fig. 4b. The TG curves of all obtained samples exhibit the typical profile expected for lignocellulosic materials [21]. The first decomposition stage (below 120 °C) corresponds to the release of physically adsorbed water, with weight losses ranging from 2 to 4% for all samples. The second stage (between 200 and 361 °C) represents the largest weight loss (50%), which is associated with the beginning of the processes of depolymerization, decomposition and rearrangement of the lignocellulose materials. Finally, the last stage (between 360 and 600 °C) corresponds the char formation around 25%.

Figure 4c shows the XRD pattern of AFPS. The diffraction profile corresponds to cellulose I, characterized by three main peaks [22]: a prominent peak at 22.2° (002), a broad peak around 15.6° (101), and a third at 34.4° (040). The crystallinity index (CI) values obtained were approximately 70% for all cellulose samples, demonstrating that the organosolv process yields cellulose with consistent CI values, indicating high reproducibility of the method.

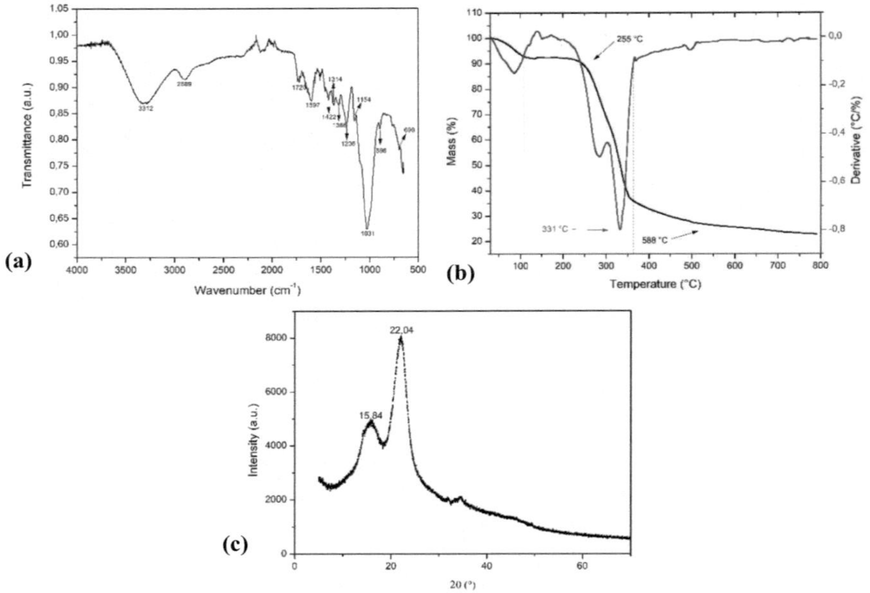

Fig. 4 **a** FTIR spectra of the AFPS. **b** TG curve recorded under nitrogen atmosphere for AFPS. **c** XRD curve of AFPS

The açaí palm stem fiber sample exhibited a crystallinity index (CI) of 50.5%. This value is comparable to that reported for *A. americana* L., which presented a CI of 66.0% using the absorbance-ratio method [23]. Lower CI values, such as 43.0%, have been reported for *A. angustifolia* [24], while Rosli et al. [7] found a CI of 59% for raw fibers of *A. angustifolia*. Therefore, it can be concluded that the crystallinity index of AFPS is within the range reported for other lignocellulosic fibers, confirming its comparable structural organization.

Conclusion

The results of the mechanical analysis indicate that fibers with a smaller diameter exhibited higher tensile strength and modulus of elasticity compared to those with a larger diameter, highlighting the potential for developing stronger composites using finer natural fibers. SEM revealed surface roughness and longitudinal deformations on the fibers. The cross-sectional analysis, which focused on the rupture area of the fiber, showed interconnected microfibrils linked by the lumen. Thermal analysis demonstrated a degradation temperature of up to 331.2 °C, which is impressive for a naturally derived material. Additionally, a crystallinity index of 50.5% confirms that the fiber's properties are compatible with lignocellulosic fibers commonly used as reinforcement in composite materials.

References

1. Silva Cedrim PCA, Barros EMA, Do Nascimento TG (2018) Antioxidant properties of acai (Euterpe oleracea) in the metabolic syndrome. Braz J Food Technol 21. Epub ahead of print 2018. https://doi.org/10.1590/1981-6723.09217
2. Laurindo LF, Barbalho SM, Araújo AC et al (2023) Açaí (Euterpe oleracea Mart.) in health and disease: a critical review. Nutrients 15. Epub ahead of print 2023. https://doi.org/10.3390/nu15040989
3. Smith N (2015) Euterpe oleracea. Geobot Stud 239–258
4. Park S, Baker JO, Himmel ME et al (2010) Cellulose crystallinity index: measurement techniques and their impact on interpreting cellulase performance. Biotechnol Biofuels 3:10
5. TAPPI T 264 cm-97 (1997) Preparation of wood for chemical analysis. Tappi Press, Atlanta
6. Wise LE, Murphy M (1946) A chlorite holocellulose, its fractionation and bearing on summative wood analysis and studies on the hemicelluloses. APA 122:35–43
7. Rosli NA, Ahmad I, Abdullah I (2013) Isolation and characterization of cellulose nanocrystals from agave angustifolia fibre. BioResources 8:1893–1908
8. García-Méndez RF, Cortés-Martínez CI, Almendárez-Camarillo A (2020) Thermochemical and tensile mechanical properties of fibers mechanically extracted from leaves of Agave angustifolia Haw. J Nat Fibers 19:3171–3185
9. Zhao H, Kwak JH, Conrad Zhang Z et al (2007) Studying cellulose fiber structure by SEM, XRD, NMR and acid hydrolysis. Carbohydr Polym 68:235–241
10. da Maradini GS, Oliveira MP, da Guanaes GMS et al (2020) Characterization of polyester nanocomposites reinforced with conifer fiber cellulose nanocrystals. Polymers (Basel) 12:1–19
11. Kong X, Zhao L, Curtis JM (2016) Polyurethane nanocomposites incorporating biobased polyols and reinforced with a low fraction of cellulose nanocrystals. Carbohydr Polym 152:487–495
12. Oprea S, Potolinca VO, Gradinariu P et al (2016) Synthesis, properties, and fungal degradation of castor-oil-based polyurethane composites with different cellulose contents. Cellulose 23:2515–2526
13. Pei A, Malho JM, Ruokolainen J et al (2011) Strong nanocomposite reinforcement effects in polyurethane elastomer with low volume fraction of cellulose nanocrystals. Macromolecules 44:4422–4427
14. Zanini NC, de Souza AG, Barbosa RFS et al (2022) Eco-friendly composites of polyurethane and sheath palm residues. J Cell Plast 58:139–158
15. El Oudiani A, Msahli S, Sakli F (2017) In-depth study of agave fiber structure using Fourier transform infrared spectroscopy. Carbohydr Polym 164:242–248
16. da Parize DDS, de Oliveira JE, Williams T et al (2017) Solution blow spun nanocomposites of poly(lactic acid)/cellulose nanocrystals from Eucalyptus kraft pulp. Carbohydr Polym 174:923–932
17. De Souza AG, Kano FS, Bonvent JJ et al (2017) Cellulose nanostructures obtained from waste paper industry: a comparison of acid and mechanical isolation methods. Mater Res 20:209–214
18. Li M, Zhao X, Li Y et al (2020) Synergistic improvement for mechanical, thermal and optical properties of PVA-co-PE nanofiber/epoxy composites with cellulose nanocrystals. Compos Sci Technol 188:107990
19. Khalid MY, Al Rashid A, Arif ZU et al (2021) Natural fiber reinforced composites: sustainable materials for emerging applications. Results Eng 11:100263
20. Stelea L, Filip I, Lisa G et al (2022) Characterisation of Hemp fibres reinforced composites using thermoplastic polymers as matrices. Polymers (Basel) 14. Epub ahead of print 2022. https://doi.org/10.3390/polym14030481

21. Benítez-Guerrero M, López-Beceiro J, Sánchez-Jiménez PE et al (2014) Comparison of thermal behavior of natural and hot-washed sisal fibers based on their main components: cellulose, xylan and lignin. TG-FTIR analysis of volatile products. Thermochim Acta 581:70–86
22. Boissou F, Muhlbauer A, De Oliveira VK et al (2014) Transition of cellulose crystalline structure in biodegradable mixtures of renewably-sourced levulinate alkyl ammonium ionic liquids, γ-valerolactone and water. Green Chem 16:2463–2471
23. Oudiani A El, Chaabouni Y, Msahli S et al (2011) Crystal transition from cellulose I to cellulose II in NaOH treated Agave americana L. fibre. Carbohydr Polym 86:1221–1229
24. Hidalgo-Reyes M, Caballero-Caballero M, HernáNdez-Gómez LH et al (2015) Chemical and morphological characterization of Agave angustifolia bagasse fibers. Bot Sci 93:807–817

Development and Characterization of Coffee Husk-Reinforced Polystyrene Biocomposites via Thermopressing

Leonardo Martins Nazário Silva, Marcela Coelho da Silva, Felipe Gabriel Santos Araújo, Felipe Perissé Duarte Lopes, Hemanuelly Ferreira Breda Lan Oliveira, Henry Alonso Colorado Lopera, and Michel Picanço Oliveira

Abstract The generation of agro-industrial residues, such as coffee husks, represents an environmental challenge due to their large volume and inadequate disposal. Incorporating these residues into polymer matrices offers an alternative route to reduce environmental impacts and promote the development of sustainable materials. In this study, polystyrene-based composites reinforced with coffee husk particles (40–70%) were produced by hot pressing and evaluated through flexural and Charpy impact tests. The results revealed a progressive reduction in flexural strength and impact resistance with increasing particle content, mainly attributed to weak interfacial adhesion, void formation, and particle agglomeration. Despite these limitations, the use of coffee husk in composite production demonstrates its potential as a renewable reinforcement, enabling high levels of residue utilization while contributing to strategies aimed at reducing plastic waste and environmental pollution.

L. M. N. Silva · M. C. da Silva · F. G. S. Araújo · H. F. B. L. Oliveira · M. P. Oliveira (✉)
Forest and Wood Sciences Department, Federal University of Espírito Santo, Jeronimo Monteiro, Brazil
e-mail: michel.oliveira@ufes.br

L. M. N. Silva
e-mail: leonardo.mn.silva@edu.ufes.br

M. C. da Silva
e-mail: marcela.c.silva@edu.ufes.br

F. G. S. Araújo
e-mail: felipe.g.araujo@edu.ufes.br

F. P. D. Lopes
Materials Science and Engineering Department, State University of Northern Rio de Janeiro, Campos dos Goytacazes, Brazil
e-mail: perisse@uenf.br

H. A. C. Lopera
CCComposites Lab, Universidad de Antioquia, Medellín, Colombia
e-mail: henry.colorado@udea.edu.co

H. A. C. Lopera et al. (eds.), *Natural Fibers and Biocomposites 2026*, The Minerals, Metals & Materials Series, https://doi.org/10.1007/978-3-032-13577-3_16

Keywords Lignocellulosic waste · Thermoplastic polymers · Biocomposites

Introduction

The search for new materials capable of meeting specific technical and functional requirements has intensified considerably in recent years. This trend is strongly associated with the targets established in international agreements, such as the Paris Agreement, which aim to mitigate climate change, as well as with the interest of companies in reducing their pollutant gas emissions to obtain carbon credits. In this context, the need to develop innovative and environmentally sustainable solutions that reconcile technological performance with lower ecological impact becomes evident.

Among such solutions, polymer composites reinforced with natural materials stand out as promising alternatives for diverse industrial applications. These materials exhibit relevant characteristics, such as suitable structural and functional properties, biodegradability, and renewability [1, 2]. Such attributes confer on these composites a more ecological and economically viable character when compared to those reinforced with synthetic fibers, such as glass or carbon, whose production costs and environmental impacts tend to be significantly higher [1].

Another factor that reinforces the relevance of using natural reinforcements is the growing consumption of plastics. According to recent data, global production of this material increased from 335 million tons in 2016 to 367 million tons in 2020, with polystyrene (PS) comprising part of this demand [3]. If effective measures are not implemented, it is estimated that by 2050 approximately 12 million tons of plastic waste may be improperly discarded into the environment [4, 5]. This scenario underscores the urgency of incorporating renewable residues into the development of new composites as a means of reducing the accumulation of plastics with limited degradability.

Polystyrene, in turn, is a C–C chain polymer characterized by high resistance to hydrolysis, which significantly delays its natural degradation. Although it presents some vulnerability to thermal oxidation, this process rarely occurs under environmental conditions, as it requires specific temperatures and situations that are not commonly found in nature. Consequently, PS exhibits high persistence in the environment, accumulating over long periods and contributing to the negative impacts of plastic pollution [6].

In parallel, coffee, one of the most consumed agricultural commodities worldwide, is notable for its composition rich in nutrients and secondary metabolites, such as caffeine, tannins, and polyphenols [7]. Brazil ranks among the three largest global producers, with a forecasted production of 51.8 million bags in 2025 to meet part of the global demand estimated at 168.1 million bags [8]. This high production volume generates large amounts of residues, among which coffee husk stands out, whose proper management represents both an environmental challenge and an opportunity for its use in higher value-added materials.

In view of the foregoing, the present study aims to develop and evaluate polystyrene composites reinforced with coffee husk particles, seeking to analyze their mechanical properties and discuss the potential use of this agricultural residue as a sustainable alternative for reducing the environmental impact associated with the use of synthetic polymers.

Materials and Methods

Materials

The coffee husk residues used in this study were supplied by Cafeeira Krohling, a company located in the state of Espírito Santo, Brazil. Initially, the material was subjected to milling and subsequently classified by sieving. Among the obtained fractions, the one retained between the 40 and 60 mesh sieves was selected, corresponding to particles with an average size of approximately 0.25 mm. For the composition of the matrix phase, polystyrene (PS) thermoplastic was employed, purchased from the company Unigel.

Methods

Fabrication of the Panels

Prior to the pressing process, the coffee husk particles were oven-dried at 100 °C for 24 h. After drying, the material was stored in a desiccator for 30 min to minimize atmospheric moisture absorption before incorporation into the polymer, which had also been previously dried. The consolidation of the composite panels was carried out using a Solab hydraulic press, model SL-12/20B. The material was subjected to a pressure of 16 tons (±4) and maintained under heating at 120 °C for 30 min. The composites were fabricated with different proportions of coffee husk particles (Fig. 1), according to the percentages presented in Table 1.

Flexural and Charpy Impact Tests

The flexural test was conducted on a universal testing machine manufactured by Oswaldo Filizola, model AME-2kN, in strict accordance with ASTM D790 [8]. The specimens were positioned in a three-point bending configuration, with the load applied at a constant strain rate, as prescribed by the standard. Data recording and processing were performed using the DynaView Standard/Pro M.3 software.

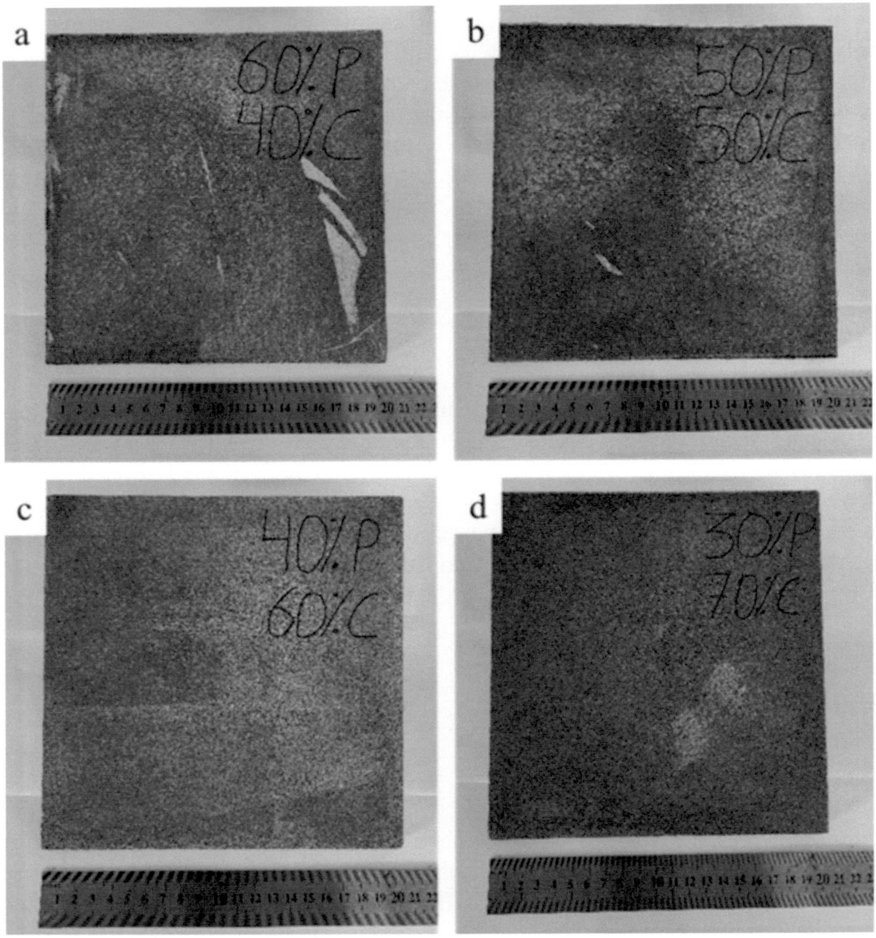

Fig. 1 Panels CH40 (**a**), CH50 (**b**), CH60 (**c**), and CH70 (**d**) after the pressing process

Table 1 Formulations of polystyrene composites reinforced with coffee husk particles

Sample	Coffee husk (%)	Polystyrene (%)
CH40	40	60
CH50	50	50
CH60	60	40
CH70	70	30

For the determination of impact toughness, the specimens were first notched at the center using a CEAST manual notching device, model Notchavas, configured to produce notches with a depth of 2.5 mm and an angle of 45°, in compliance with ASTM D6110 [9]. Subsequently, the impact tests were carried out on an XJ Series Impact Testing Machine, operating with a pendulum hammer of 11 J and an impact velocity of 2.9 m/s. It is noteworthy that all mechanical tests were performed with 24 replicates for each formulation, containing coffee husk particle contents of 40, 50, 60, and 70%.

Results and Discussion

The analysis of Fig. 2 shows that the progressive increase in the content of coffee husk particles incorporated into the polystyrene matrix, ranging from 40 to 70%, resulted in a significant reduction in the maximum flexural strength. This behavior is directly associated with the low interfacial compatibility between the hydrophobic polymer matrix and the predominantly hydrophilic lignocellulosic particles, which compromises the fiber–matrix adhesion and, consequently, the stress transfer. The absence of efficient interaction favors particle detachment and the formation of regions susceptible to failure nucleation under mechanical loading [10]. In the same regard, Jaramillo et al. [11] and Melyna and Afridana [12] emphasize that the increase in coffee husk fraction in polymer composites leads to a reduction in mechanical properties, a phenomenon attributed to the presence of voids, heterogeneity in particle dispersion, and, above all, the absence of coupling agents. According to the authors, such factors, combined with the high hydrophilicity of the particles, hinder interfacial adhesion, promote moisture absorption, and accelerate interface degradation, thereby intensifying crack propagation in the composite.

Complementarily, Joseph et al. [13] highlight that the lower flexural strength may also be related to the irregular shape and random distribution of fibers within the matrix. This structural arrangement interrupts the continuity of the polymer–particle bonding and, in cases of fibers reduced in size, accentuates the tendency to agglomeration, making the dispersion pattern a decisive factor for the final mechanical performance of the composite.

Figure 3 shows that the Charpy impact resistance decreases progressively with the increase in coffee husk particle content from 40 to 70% in the polystyrene matrix. This trend is attributed to the weak interfacial adhesion between the hydrophobic polymer matrix and the hydrophilic lignocellulosic particles, which compromises efficient stress transfer and energy dissipation under dynamic loading. As reported by Yang et al. [14], the reduction in impact energy is a recurrent phenomenon in natural composites, resulting from the presence of fragile zones and the accelerated propagation of cracks originating from interfacial defects.

Similar results were found by Nunes et al. [15], who observed a decrease in toughness in polymer composites reinforced with plant fibers, associating such behavior with the presence of voids, poor dispersion, and particle detachment from the matrix

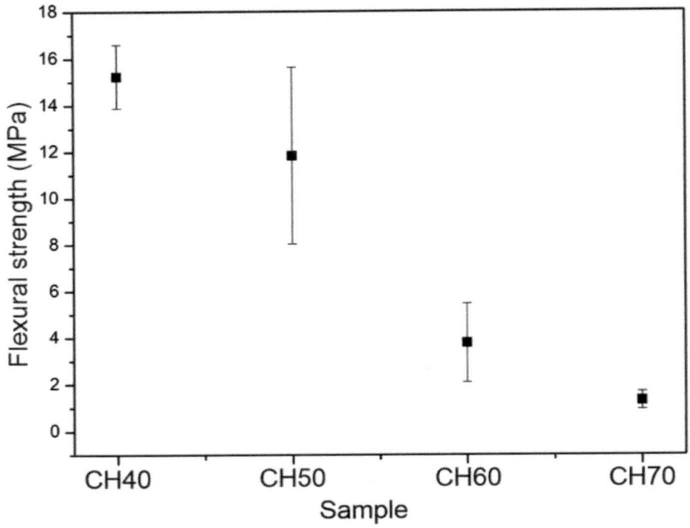

Fig. 2 Effect of increasing coffee husk particle content (40–70%) on the flexural strength of polystyrene composites

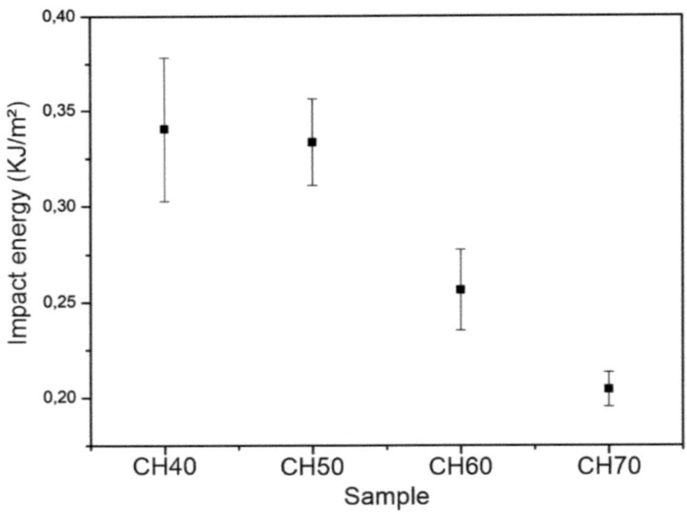

Fig. 3 Effect of increasing coffee husk particle content (40–70%) on the Charpy impact resistance of polystyrene composites

during impact. These microstructural aspects act as stress concentration points, reducing the material's capacity to absorb energy.

Convergently, Nourbakhsh et al. [16] highlighted that, although increasing the lignocellulosic reinforcement fraction may improve stiffness, there is a significant compromise in impact resistance when coupling agents are not employed. This effect arises from the formation of interfacial gaps and particle agglomeration at higher filler contents, which reduce the structural integrity of the composite. In the same direction, Melyna and Afridana [12] emphasized that residues such as coffee husk exhibit high hydrophilicity, a characteristic that hinders adhesion to the thermoplastic matrix and intensifies moisture absorption. This condition enhances interface degradation and promotes crack nucleation, resulting in a marked reduction in impact resistance.

Conclusion

The incorporation of coffee husk particles into the polystyrene matrix enabled the production of biocomposites with high residue utilization, reaching up to 70%. However, the progressive increase in particle content resulted in a significant reduction in flexural strength and impact resistance, primarily due to weak interfacial adhesion and particle agglomeration. Despite these limitations, the results demonstrate the potential of coffee husk as a sustainable reinforcement for polymer composites, reinforcing its relevance in strategies aimed at reducing the environmental impact of plastic materials.

Acknowledgements This study was financed in part by the Espírito Santo Research and Innovation Support Foundation (FAPES), Brazil, through EDITAL FAPES Nº 03/2021 - UNIVERSAL (grant numbers: 456/2021); EDITAL FAPES Nº 28/2022 – UNIVERSAL (grant numbers: 932/2023) and National Council for Scientific and Technological Development – CNPq. Additionally, the Coordenação de Aperfeiçoamento de Pessoal de Nível Superior (CAPES) in Brazil provided funding under Finance Code 001.

References

1. Ajayi NE, Rusnakova S, Ajayi AE, Ogunleye RO, Agu SO, Amenaghawon AN (2025) A comprehensive review of natural fiber reinforced Polymer composites as emerging materials for sustainable applications. Appl Mater Today 43:102666. https://doi.org/10.1016/j.apmt.2025.102666
2. Chung DDL (2019) A review of multifunctional polymer-matrix structural composites. Compos B Eng 160:644–660. https://doi.org/10.1016/j.compositesb.2018.12.117
3. EUROPE P (2021) Plastics Europe Association of Plastics Manufacturers Plastics—the facts 2021 an analysis of European plastics production, demand and waste data
4. Dong D, Guo Z, Yang X, Dai Y (2024) Comprehensive understanding of the aging and biodegradation of polystyrene-based plastics. Environ Pollut 342:123034. https://doi.org/10.1016/j.envpol.2023.123034

5. Geyer R, Jambeck JR, Law KL (2017) Production, use, and fate of all plastics ever made. Sci Adv 3. https://doi.org/10.1126/sciadv.1700782
6. Anshu K, Kenttämaa HI, Thengane SK (2024) A comprehensive review on co-pyrolysis of lignocellulosic biomass and polystyrene. Renew Sustain Energy Rev 205:114832. https://doi.org/10.1016/j.rser.2024.114832
7. Fernandes G (2007) Extração e purificação de cafeína da casca de café. Universidade Federal de Uberlândia, Dissertação (Mestrado em Engenharia Química)
8. de A. CONAB CN (2024) Acompanhamento da safra brasileira de café, 1st ed. Brasília
9. ASTM D790 (2017) Standard test methods for flexural properties of unreinforced and reinforced plastics and electrical insulating materials. American Society for Testing and Materials D790–17 Stand
10. ASTM D6110 (2018) Standard test method for determining the charpy impact resistance of notched specimens of plastics. American Society for Testing and Materials D6110–18 Stand 1–17
11. Huang L, Mu B, Yi X, Li S, Wang Q (2018) Sustainable use of coffee husks for reinforcing polyethylene composites. J Polym Environ 26:48–58. https://doi.org/10.1007/s10924-016-0917-x
12. Jaramillo LY, Vásquez-Rendón M, Upegui S, Posada JC, Romero-Sáez M (2021) Polyethylene-coffee husk eco-composites for production of value-added consumer products. Sustain Environ Res 31:34. https://doi.org/10.1186/s42834-021-00107-6
13. Melyna E, Afridana AP (2023) The effect of coffee husk waste addition with alkalisation treatment on the mechanical properties of polypropylene composites. Equilib J Chem Eng 7(14). https://doi.org/10.20961/equilibrium.v7i1.68556.
14. Joseph S (2002) A comparison of the mechanical properties of phenol formaldehyde composites reinforced with banana fibres and glass fibres. Compos Sci Technol 62:1857–1868. https://doi.org/10.1016/S0266-3538(02)00098-2
15. Yang H-S, Kim H-J, Park H-J, Lee B-J, Hwang T-S (2007) Effect of compatibilizing agents on rice-husk flour reinforced polypropylene composites. Compos Struct 77:45–55. https://doi.org/10.1016/j.compstruct.2005.06.005
16. Nunes SG, da Silva LV, Amico SC, Viana JD, Amado FDR (2016) Study of composites produced with recovered polypropylene and piassava fiber. Mater Res 20:144–150. https://doi.org/10.1590/1980-5373-mr-2016-0659
17. Nourbakhsh A, Baghlani FF, Ashori A (2011) Nano-SiO_2 filled rice husk/polypropylene composites: physico-mechanical properties. Ind Crops Prod 33:183–187. https://doi.org/10.1016/j.indcrop.2010.10.010

Development and Processing of Thermoplastic Matrix Reinforced with Continuous Banana Fiber via Fused Filament Fabrication

Oswaldo Rivero-Romero, Jimy Unfried-Silgado, and Henry A. Colorado

Abstract Banana crops generate a considerable number of lignocellulose-rich residues. One strategy for managing these residues is to use additive manufacturing technologies, which can increase their aggregate value. In this study, a fused filament fabrication (FFF) printer was modified for 3D printing continuous banana fiber yarns (CBFY) in a polylactic acid (PLA) matrix. For this purpose, a new 3D print head with in-situ impregnation was developed to process the printed parts. The CBFY were treated with mercerization and a polymer coating. The most relevant problems related to depositing the printing material made from CBFY and the PLA matrix were discussed. Additionally, tensile and flexural testing were used to evaluate the mechanical properties of the parts. The properties of the printed parts were analyzed using analysis of variance (ANOVA).

Keywords Fused filament fabrication · Continuous natural fiber · Polymer coating · Mechanical properties

Introduction

3D printing, or additive manufacturing (AM), enables rapid prototyping of complex geometries [1]. Fused filament fabrication (FFF) is a popular AM technology for thermoplastics like PLA, ABS, and Nylon [2, 3]. However, these materials have limited load capacity for functional components [4]. Continuous fiber-reinforced composites (CFRCs) address this limitation and are used in aerospace, automotive, and energy sectors [5]. FFF for thermoplastic composites often uses pre-impregnated

O. Rivero-Romero (✉) · H. A. Colorado
Mechanical Engineering Department, University of Cordoba, Research Group ICT, Monteria, Cordoba, Colombia
e-mail: oswaldo.rivero@udea.edu.co

O. Rivero-Romero · J. Unfried-Silgado
Mechanical Engineering Department, University of Antioquia, Research Group CCComposite, Medellin, Antioquia, Colombia

H. A. C. Lopera et al. (eds.), *Natural Fibers and Biocomposites 2026*, The Minerals, Metals & Materials Series, https://doi.org/10.1007/978-3-032-13577-3_17

filament or in-situ impregnation, where fiber and polymer are combined in the print head [6, 7]. While synthetic fibers like carbon or glass are common, their production relies on non-renewable resources and generates waste [10, 11]. This has spurred interest in sustainable biocomposites aligned with circular economy principles [12].

The development of continuous natural fiber-reinforced composites (CNFRCs) via in-situ FFF is emerging. Studies using jute, flax, and ramie fibers have shown significant mechanical improvements [5, 14, 15]. However, printing parameters significantly affect the interfacial and mechanical properties of CNFRCs [16, 17]. Utilizing natural fibers from agro-industrial waste, such as banana fiber, offers an eco-friendly alternative with good mechanical potential [19, 20]. Banana fiber, extracted from the pseudo-stem, has a tensile strength of 600–800 MPa and modulus of 29–32 GPa [23], making it a promising reinforcement.

This study investigates the development and processing of a PLA matrix reinforced with continuous banana fiber via FFF. The influence of printing parameters on printing quality was evaluated. Additionally, the mechanical properties of composites with varying fiber content and fiber orientations were also analyzed using ANOVA.

Method and Materials

Materials

Banana fibers were extracted from the pseudo-stem of Harton variety plants (Cereté, Colombia). Sheaths were cut into 1.5 m pieces, and fibers were manually decorticated, washed, and dried at 40 °C for 4 h. Figure 1 shows the process carried out to obtain banana fiber. The matrix material was commercial eSUN® PLA filament (Ø1.75 mm). To create yarn for printing, fibers of similar diameter were selected and joined end-to-end with PLA to form a continuous single-fiber yarn (SBF). Four SBFs were then twisted into a banana fiber yarn (BFY) using a spinning wheel. The fibers for BFY were treated in a 5 wt.% NaOH solution for 1 h at room temperature, neutralized, rinsed, and dried. The BFY was then coated with flexible epoxy resin and cured to produce coated banana fiber yarn (CBFY).

Printer Modification and Printing Process

A Cartesian FFF 3D printer (build volume 220 × 220 × 210 mm^3) was modified with a new print head for in situ impregnation (see Fig. 2). This head simultaneously fed the continuous banana fiber and PLA filament, combining them at the hotend where the polymer melts before extrusion, as seen in Fig. 2b.

Fig. 1 Process carried out to obtain banana fibers

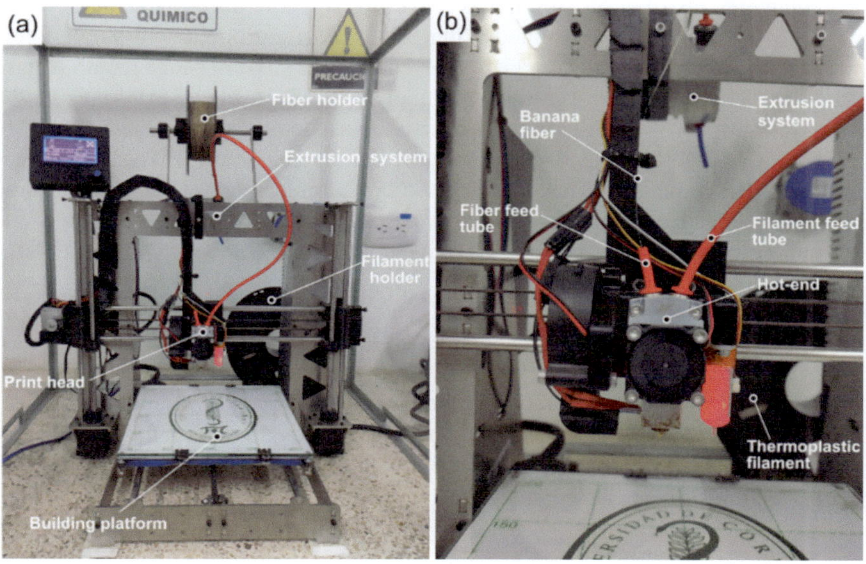

Fig. 2 **a** Overview of the 3D printer and **b** new print head used to produce 3D-printed composite

Specimen Preparation

Single-trip experiments were first conducted to determine printability, evaluating parameters like printing speed, layer height, and extrusion width. Optimal conditions were defined to prevent fiber damage, ensure adhesion, and achieve the intended strand shape. On the other hand, a 3 × 2 factorial design was used for mechanical testing (see Table 1), evaluating two fiber orientations (0° and 45°) and two fiber contents. Tensile (ASTM D638) and flexural (ASTM D790) specimens were printed with four replicates each (48 total specimens). Printing parameters were constant:

Table 1 3 × 2 Factorial experiment design

Orientation (B)	Content fiber (A)	
	1.0% (A1)	3.98% (A2)
0° (B1)	A1B1	A2B1
45° (B2)	A1B2	A2B2

layer thickness (1 mm), extrusion width (2.4 mm), nozzle temperature (200 °C), print speed (300 mm/min), and bed temperature (60 °C).

Characterization and Mechanical Testing

Fiber diameter distribution was measured using ImageJ® on optical microscopy images. Morphology was examined using SEM. Tensile properties of single fibers and BFY were tested with a Shimadzu® universal tester (50 mm gauge length, 2 mm/min speed). Thermal stability of fibers was analyzed by TGA (25 °C to 940 °C, 15 °C/min, N_2 atmosphere). PLA thermal properties were determined by DSC (25 °C to 500 °C, 5 °C/min, N_2) to identify appropriate printing temperatures. The microstructure of printed composites was observed via optical microscopy. The fiber volume fraction (V_f) was calculated using Eq. 1, where S_f is the cross-sectional area of the fiber and S_C is the cross-sectional area of the composite.

$$V_f = \frac{S_f}{S_C} \times 100\% \quad (1)$$

Tensile and flexural tests were performed on a Shimadzu® Autograph AG–X. Tensile tests used a 10 kN load cell, 25 mm gauge length, and 5 mm/min speed. Flexural tests used a three-point bend fixture with a 57.5 mm span and 1.53 mm/min speed. Fracture surfaces were analyzed using SEM.

Results and Discussion

Materials Characterization

Figure 3a and b show the results of the fiber diameter of the extradited fiber, selected banana fiber (SBF), and banana fiber yarn (BFY). The average diameter of extracted fibers was 251.69 ± 74.70 μm. Selected fibers for SBF had a smaller, more uniform average diameter of 187.68 ± 55.46 μm. The BFY showed a uniform diameter distribution with an average of 348.59 ± 62.27 μm. SEM revealed surface impurities on raw fibers and a helical structure with some defects in the BFY due to twisting (see Fig. 4c and d).

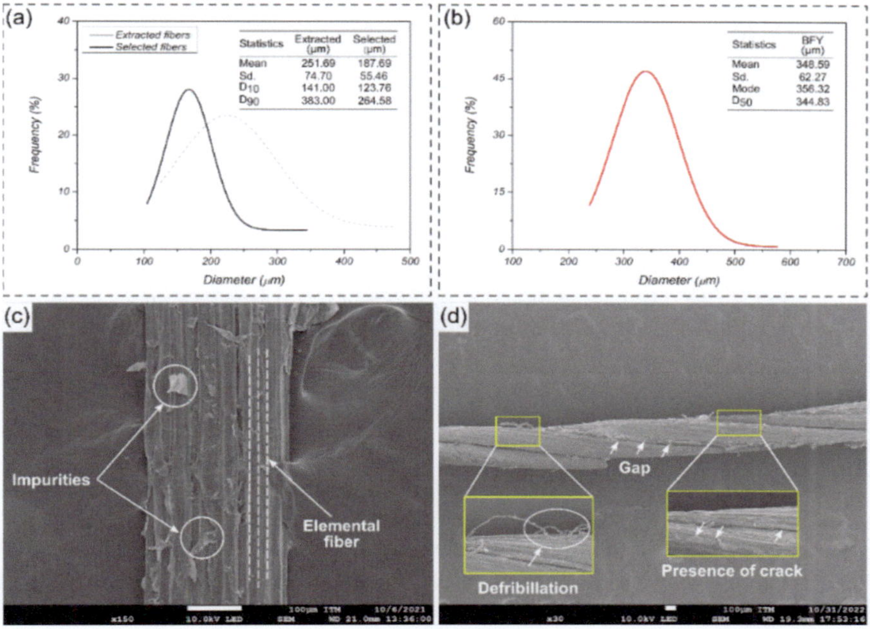

Fig. 3 Distribution of the fiber diameter: **a** extracted fibers and **b** selected fibers, and SEM micrograph: **c** banana fiber and **d** BFY

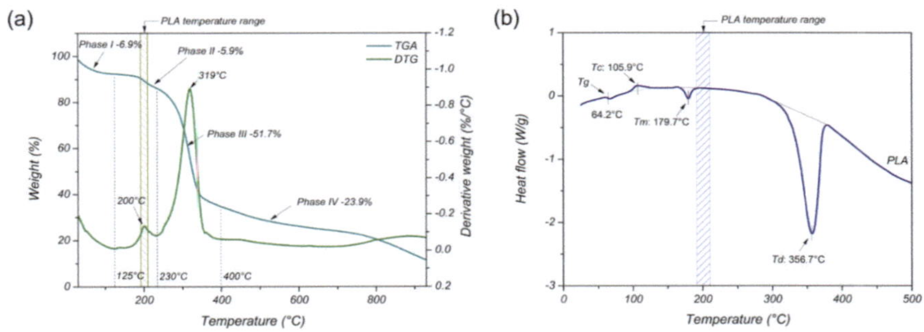

Fig. 4 **a** Thermogravimetric analysis of banana fiber and **b** differential scanning calorimetry curves of the PLA filament

Table 2 shows the tensile testing results of banana fibers. Tensile tests showed that the single banana fiber had higher strength (247.5 ± 162.1 MPa) and modulus (17.9 ± 15.1 GPa) than the BFY (94.42 ± 23.6 MPa and 5.37 ± 1.2 GPa), attributed to the weakening effect of twisting.

Figure 4 shows the thermogravimetric analysis results for banana fiber and the PLA matrix. TGA showed banana fiber degradation stages: moisture loss (<125 °C),

Table 2 Tensile properties of banana fiber and BFY

Materials	Tensile strength (MPa)	Tensile modulus (GPa)	Elongation at break (%)
Banana fiber	247.5 ± 162.1	17.9 ± 15.1	1.8 ± 0.3
BFY	94.42 ± 23.6	5.37 ± 1.2	2.70 ± 1.1

removal of extractives (125–230 °C), decomposition of hemicellulose/cellulose (230–400 °C), and lignin combustion (>400 °C). DSC of PLA identified a glass transition (Tg ~ 64.9 °C), crystallization exotherm (100–125 °C), melting point (Tm ~ 179.7 °C), and degradation temperature (~356.7 °C). The results confirmed that printing at 200 °C would melt the PLA without degrading the fiber.

Printability of Composite

Printability was assessed through single-strand experiments (Table 3). Optimal printing was achieved at a layer height of 1 mm, extrusion width of 2.4 mm, and print speed of 300 mm/min. Lower widths caused poor adhesion or fiber damage, while higher widths resulted in unintended strand shapes.

Tensile and Flexural Properties

Figure (a) shows the results of fiber volume fraction in different fiber orientations (0°, 45° and 90°) for SBF. The fiber volume fraction was consistent across different orientations (~1% for 3 layers, ~0.6% for 2 layers), confirming uniform deposition. This can also be observed in Fig. 5b. For the composite with BFY, the fiber content was 3.98%.

Figure 6 shows the composite's mechanical properties. For 0° orientation, the tensile strength decreased compared to neat PLA (2.74% for SBF and 6.05% for twisted fibers/CBFY), despite a higher fiber content (3.98%). This was attributed to poor impregnation and the helical structure of twisted fibers hindering stress transfer. For 45° orientation, data variability made it difficult to conclusively attribute changes to fiber content, though strength was generally lower than at 0° due to off-axis loading. Flexural strength for 0° samples showed decreases of 8.97% (SBF) and 9.78% (BFY) compared to neat PLA, with no significant difference between the two reinforced groups. This suggests poor fiber-matrix adhesion under bending stress. For 45° orientation, flexural strength values (70.07 ± 5.17 MPa for SBF and 75.21 ± 2.46 MPa for CBFY) were not significantly different from neat PLA, likely due to poor interlayer adhesion dominating the mechanical response.

Table 3 Single trip printability of banana fiber-impregnated thermoplastic material using the FFF process with in-situ impregnation

Printing speed (mm/min)	Layer thickness (mm)	Extrusion width (mm)	Printability
200	0.6	1	Not attached
		1.6	Not attached
		2	Not attached
		2.4	Not attached
		2.8	Damaged
	0.8	1	Not attached
		1.6	Not attached
		2	Damaged
		2.4	Damaged
		2.8	Damaged
	1	1	Not attached
		1.6	Damaged
		2	Damaged
		2.4	Printable
		2.8	Unintended shape
300	0.6	1	Not attached
		1.6	Not attached
		2	Not attached
		2.4	Not attached
		2.8	Damaged
	0.8	1	Not attached
		1.6	Not attached
		2	Damaged
		2.4	Damaged
		2.8	Damaged
	1	1	Damaged
		1.6	Damaged
		2	Damaged
		2.4	Printable
		2.8	Unintended shape

Anova

Table 4 shows the results of the ANOVA for tensile strength. According to the P-value column, all of which are less than 0.05, the effects of content, orientation, and their interaction are significant on the tensile strength of the manufactured materials, contributing 5.31%, 96.84%, and 1.28%, respectively. Qualitatively, these results

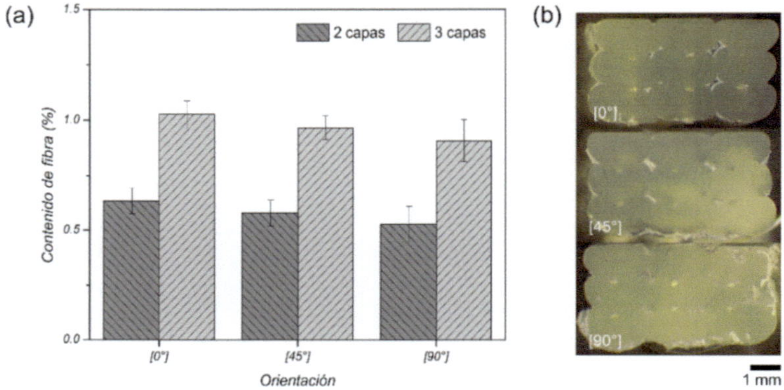

Fig. 5 **a** Fraction volume fiber with fiber orientations of 0°, 45°, and 90°; **b** optical images of the cross-sections of rectangular samples

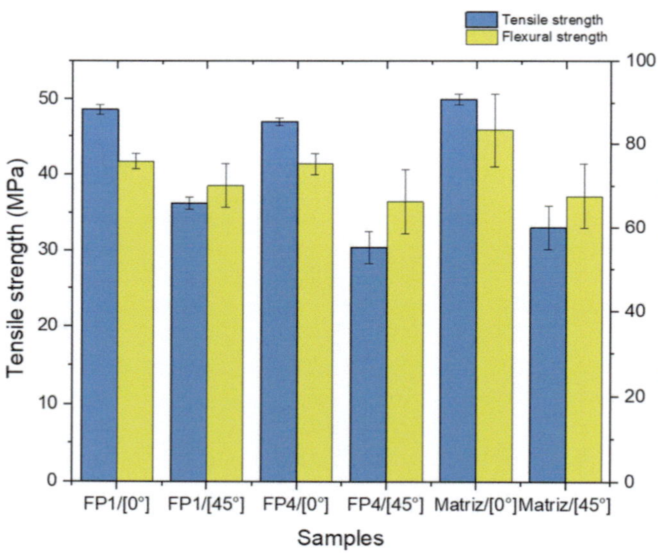

Fig. 6 Tensile strength and flexural strength of PLA/banana composite

indicate that the content and orientation of the reinforcing fiber, as well as their interaction, influence the tensile strength of the composite material developed. On the other hand, Table 4 shows the results of the ANOVA for flexural strength. These results show that only orientation had a significant effect on the flexural strength of the obtained composite material, with a contribution of 40.69%. In contrast, the content and the interaction of both factors showed no statistical influence on the variation of the response, according to the obtained P-value (Table 5).

Table 4 Results of the ANOVA for tensile strength

Source	Sum of squares	df	Mean square	F-ratio	P-value	Contribution (%)	Observation
A: Content	39.45	1	39.45	85.89	0	5.31	Significant
B: Orientation	719.69	1	719.69	1566.9	0	96.84	Significant
AB	9.55	1	9.55	20.78	0.0008	1.28	Significant
Error	5.05	11	0.46				
Total	743.17	14					

Table 5 Results of the ANOVA for flexural strength

Source	Sum of squares	df	Mean square	F-ratio	P-value	Contribution (%)	Observation
A: Content	20.23	1	20.23	0.84	0.38	3.77	Not significant
B: Orientation	218.42	1	218.42	9.1	0.01	40.69	Significant
AB	9.9	1	9.9	0.41	0.53	1.84	Not significant
Error	288.19	12	24.02				
Total	536.75	15					

Conclusions

This study successfully developed a process for 3D printing continuous banana fiber-reinforced PLA composites using a modified FFF printer with in-situ impregnation. Mercerization and epoxy coating were used to treat the banana fiber yarns.

- Optimal printing parameters were identified as a layer height of 1 mm, extrusion width of 2.4 mm, and print speed of 300 mm/min to achieve viable strands without fiber damage.
- The thermal analysis confirmed that printing at 200 °C is suitable, melting the PLA without degrading the banana fibers.
- The incorporation of continuous banana fiber, particularly in twisted yarn form, did not lead to the expected improvement in tensile and flexural properties compared to neat PLA. This was primarily attributed to challenges in achieving effective impregnation and strong interfacial adhesion between the fiber and the matrix, as well as the potential weakening effect of twisting the fibers into a yarn.
- The 0° fiber orientation generally yielded better mechanical properties than the 45° orientation, consistent with the anisotropic nature of fiber-reinforced composites.

- The ANOVA results demonstrate that fiber orientation is the dominant factor influencing the mechanical properties of the 3D-printed composites, with a particularly strong effect on tensile strength (96.84% contribution). While fiber content and its interaction with orientation also affect tensile performance, only orientation significantly impacts flexural strength. These findings highlight the critical importance of print path optimization for maximizing mechanical performance in continuous fiber-reinforced composites, whereas increasing fiber content alone shows limited benefits for flexural properties.

Future work should focus on optimizing fiber treatment and printing parameters to enhance interfacial adhesion and maximize the potential of banana fiber as a sustainable reinforcement in additive manufacturing.

Acknowledgements The authors thank the Colombian Ministry of Science, Technology, and Innovation for funding through grant 891-2020. Thanks to the Grupo de Materiales Avanzados y Energía-MATyER and the University of Córdoba for facility access and support (project FI-06-19).

References

1. Van Der Klift F, Koga Y, Todoroki A, Ueda M, Hirano Y, Matsuzaki R (2016) 3D Printing of continuous carbon fibre reinforced thermo-plastic (CFRTP) tensile test specimens. Open J Compos Mater 06(01):18–27. https://doi.org/10.4236/ojcm.2016.61003
2. Parandoush P, Lin D (2017) A review on additive manufacturing of polymer-fiber composites. Composite structures, vol 182. Elsevier Ltd, pp 36–53. https://doi.org/10.1016/j.compstruct.2017.08.088
3. Zhang X, Chen L, Mulholland T, Osswald TA (2019) Effects of raster angle on the mechanical properties of PLA and Al/PLA composite part produced by fused deposition modeling. Polym Adv Technol 30(8):2122–2135. https://doi.org/10.1002/pat.4645
4. Ferreira I, Machado M, Alves F, Torres Marques A (2019) A review on fibre reinforced composite printing via FFF. Rapid Prototyp J 25(6). Emerald Group Holdings Ltd., pp 972–988. https://doi.org/10.1108/RPJ-01-2019-0004
5. Terekhina S, Egorov S, Tarasova T, Skornyakov I, Guillaumat L, Hattali ML (2022) In-nozzle impregnation of continuous textile flax fiber/polyamide 6 composite during FFF process. Compos Part A Appl Sci Manuf 153. https://doi.org/10.1016/j.compositesa.2021.106725
6. Tian X et al (2022) 3D printing of continuous fiber reinforced polymer composites: development, application, and prospective. Chin J Mech Eng Addit Manuf Front 1(1):100016. https://doi.org/10.1016/j.cjmeam.2022.100016
7. Krajangsawasdi N, Blok LG, Hamerton I, Longana ML, Woods BKS, Ivanov DS (2021) Fused deposition modelling of fibre reinforced polymer composites: a parametric review. J Compos Sci 5(1). MDPI AG. https://doi.org/10.3390/jcs5010029
8. Li S, Wang K, Zhu W, Peng Y, Ahzi S, Chinesta F Science arts & métiers (SAM) contributions of interfaces on the mechanical behavior of 3D printed continuous fiber reinforced composites. https://sam.ensam.eu
9. Mashayekhi F, Bardon J, Berthé V, Perrin H, Westermann S, Addiego F (2021) Fused filament fabrication of polymers and continuous fiber-reinforced polymer composites: advances in structure optimization and health monitoring. Polymers 13(5):1–29. MDPI AG. https://doi.org/10.3390/polym13050789

10. Wu J, Gao X, Wu Y, Wang Y, Nguyen TT, Guo M (2023) Recycling carbon fiber from carbon fiber-reinforced polymer and its reuse in photocatalysis: a review. Polymers 15(1). MDPI. https://doi.org/10.3390/polym15010170
11. Stefan DS, Bosomoiu M, Stefan M (2022) Methods for natural and synthetic polymers recovery from textile waste. Polymers 14(19). MDPI. https://doi.org/10.3390/polym14193939
12. Colorado HA, Velásquez EIG, Monteiro SN (2020) Sustainability of additive manufacturing: the circular economy of materials and environmental perspectives. J Market Res 9(4):8221–8234. https://doi.org/10.1016/j.jmrt.2020.04.062
13. Le Duigou A, Chabaud G, Matsuzaki R, Castro M (2020) Tailoring the mechanical properties of 3D-printed continuous flax/PLA biocomposites by controlling the slicing parameters. Compos B Eng 203. https://doi.org/10.1016/j.compositesb.2020.108474
14. Matsuzaki R et al (2016) Three-dimensional printing of continuous-fiber composites by in-nozzle impregnation. Sci Rep 6. https://doi.org/10.1038/srep23058
15. Kuschmitz S, Schirp A, Busse J, Watschke H, Schirp C, Vietor T (2021) Development and processing of continuous flax and carbon fiber-reinforced thermoplastic composites by a modified material extrusion process. Materials 14(9). https://doi.org/10.3390/ma14092332
16. Cheng P et al (2021) Interfacial and mechanical properties of continuous ramie fiber reinforced biocomposites fabricated by in-situ impregnated 3D printing. Ind Crops Prod 170. https://doi.org/10.1016/j.indcrop.2021.113760
17. Cai R, Wen W, Wang K, Peng Y, Ahzi S, Chinesta F (2022) Tailoring interfacial properties of 3D-printed continuous natural fiber reinforced polypropylene composites through parameter optimization using machine learning methods. Mater Today Commun 32. https://doi.org/10.1016/j.mtcomm.2022.103985
18. Zhang Z, Long Y, Yang Z, Fu K, Li Y (2022) An investigation into printing pressure of 3D printed continuous carbon fiber reinforced composites. Compos Part A Appl Sci Manuf 162. https://doi.org/10.1016/j.compositesa.2022.107162
19. Awad S et al (2023) Evaluation of characterisation efficiency of natural fibre-reinforced polylactic acid biocomposites for 3D printing applications. Sustain Mater Technol 36. https://doi.org/10.1016/j.susmat.2023.e00620
20. Patel R, Dhimmar V, Kagzi S, Patel M (2022) Investigation of fused deposition modelling process parameters in 3D printing for composite material (Poly Lactic Acid and Banana Fibre). Int J Automot Mech Eng 19(3). https://doi.org/10.15282/ijame.19.3.2022.14.0774
21. Suteja J, Firmanto H, Soesanti A, Christian C (2022) Properties investigation of 3D printed continuous pineapple leaf fiber-reinforced PLA composite. J Thermoplast Compos Mater 35(11):2052–2061. https://doi.org/10.1177/0892705720945371
22. Jayaprakash K, Osama A, Rajagopal R, Goyette B, Karthikeyan OP (2022) Agriculture waste biomass repurposed into natural fibers: a circular bioeconomy perspective. Bioengineering 9(7). https://doi.org/10.3390/bioengineering9070296
23. Badanayak P, Jose S, Bose G (2023) Banana pseudostem fiber: a critical review on fiber extraction, characterization, and surface modification. J Nat Fibers 20(1). Taylor and Francis Ltd. https://doi.org/10.1080/15440478.2023.2168821

Development of Anti-corrosive Coatings Reinforced with Glass Waste

Darcy Lucas da Rocha Oliveira, David Coverdale Rangel Velasco, Victor Paes Dias Gonçalves, Michel Picanço Oliveira, Noan Tonini Simonassi, Carlos Maurício Fontes Vieira, and Felipe Perissé Duarte Lopes

Abstract Corrosion remains a critical issue in industrial environments, leading to severe structural degradation and significant economic losses, especially in sectors such as offshore oil platforms and pipelines exposed to harsh, corrosive conditions. Protective coatings, including structural paints and pipe linings, are commonly used to extend the service life of metallic structures. Epoxy resin stands out for its excellent chemical resistance, mechanical strength, and cost-effectiveness; however, its non-recyclable nature poses sustainability challenges. To address this, the present study investigates an eco-friendly composite developed by incorporating fine glass waste particles into the epoxy matrix. This approach aims to reduce the environmental footprint of conventional coatings while maintaining or improving their protective performance. Incorporation of glass waste up to 20 vol% did not compromise the mechanical properties, confirming the feasibility of this sustainable alternative. Overall, the proposed strategy promotes resource efficiency and aligns with circular economy and green engineering principles.

Keywords Sustainability · Composites · Glass waste · Coatings

D. L. da Rocha Oliveira (✉) · D. C. R. Velasco · V. P. D. Gonçalves · N. T. Simonassi · C. M. F. Vieira · F. P. D. Lopes
UENF–State University of the Rio de Janeiro, LAMAV–Advanced Materials Laboratory, Campos dos Goytacazes, Rio de Janeiro, Brazil
e-mail: darcylucas.rocha@outlook.com

D. C. R. Velasco
IFF Cabo Frio–Fluminense Federal Institute, Cabo Frio Campus, Cabo Frio, Rio de Janeiro, Brazil

M. P. Oliveira
UFES–Federal University of Espírito Santo, Jerônimo Monteiro Campus, Centro, Jerônimo Monteiro, Espírito Santo, Brazil

H. A. C. Lopera et al. (eds.), *Natural Fibers and Biocomposites 2026*, The Minerals, Metals & Materials Series, https://doi.org/10.1007/978-3-032-13577-3_18

Introduction

The growing urgency of environmental challenges, such as the increasing frequency of natural disasters, polar ice cap melting, and rising ocean acidity [3], has made sustainability a central focus in both academic and industrial contexts [1]. Consequently, there is a pressing demand for more environmentally responsible production practices, waste reuse, and the development of less harmful technologies [2].

In this context, the circular economy model has gained prominence for promoting resource efficiency through material reuse within the production cycle [4], often transforming waste into a viable and sustainable raw material [5]. This need for sustainable strategies is especially critical in highly corrosive industrial environments, such as those encountered in the oil, gas, and nuclear energy sectors [6].

Although these industries are advancing toward cleaner technologies, the structural integrity of their infrastructure remains constantly threatened by corrosion. This challenge requires the extensive use of steel and other metals whose production is carbon-intensive, resulting in substantial CO_2 emissions and material waste [6]. To mitigate structural degradation and reduce material consumption, anti-corrosive coatings are widely applied [7]. These protective layers play a crucial role in safeguarding metallic components and pipelines exposed to aggressive environments [8], particularly in offshore operations.

However, a significant drawback of these coatings is their limited environmental sustainability. Most rely on non-recyclable polymers such as epoxy resin, which, despite its low cost and excellent chemical resistance, raises concerns regarding end of life disposal [9]. Additionally, conventional performance additives such as thickeners and reinforcements are predominantly synthetic [10], further exacerbating sustainability challenges.

A promising alternative involves the incorporation of natural materials or industrial waste as dispersed phases in coating formulations [11]. Several studies have shown that adding glass particles to epoxy matrices enhances substrate adhesion [12] and can improve mechanical properties [13]. This strategy not only enhances material performance but also reduces the consumption of non-recyclable resin, contributing to a more sustainable solution.

With this in mind, the present study aims to characterize the mechanical properties of epoxy-based materials designed for anticorrosive coatings incorporating glass waste as a dispersed phase. The objective is to develop a cost-effective and environmentally friendly coating that also contributes to waste reduction by diverting glass residues from landfills [14].

Materials and Methods

The epoxy resin system employed in this study consisted of Diglycidyl Ether of Bisphenol A (DGEBA) as the base resin and Triethylenetetramine (TETA) as the curing agent. These materials, commercially designated as SQ2050 (DGEBA) and SQ3131 (TETA), respectively, were supplied by Avipol. An image of the products is presented below (Image 1).

Image 1 Resin and hardener utilized in this work

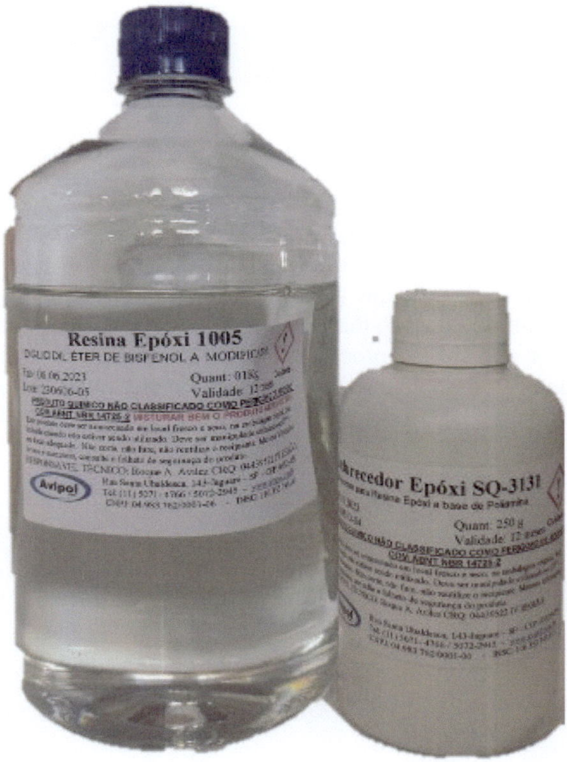

The second primary material used in this study was glass waste, as illustrated in the image below. To enable its incorporation into the epoxy matrix, the collected glass containers were processed in a ball mill to reduce particle size and obtain a finer, more uniform morphology. The images below show the material in its initial state (as received) and after undergoing the crushing and sieving processes (Image 2).

Image 2 Material before crushing and comminution (left) and after processing

The processed glass waste was sieved using a 100 mesh (0.149 mm) sieve to obtain particles with a uniform size distribution. This specific particle size was selected because it provides a smaller aspect ratio, improving workability and facilitating incorporation into the epoxy matrix for composite applications. After the comminution and sieving processes, the resulting particles were characterized, and an SEM image of the material used in the preparation of the test specimens is shown below (Image 3).

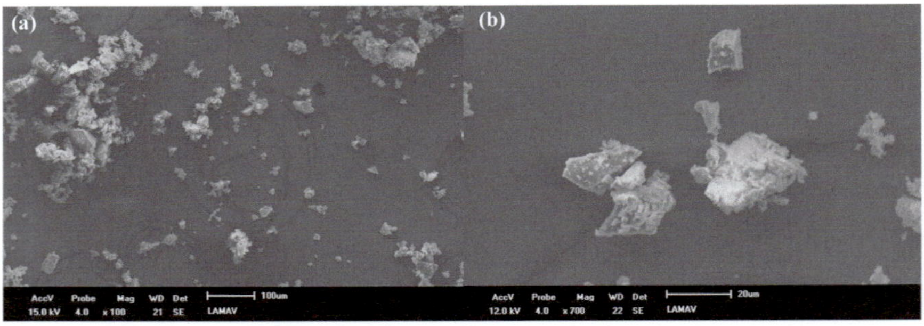

Image 3 SEM of the particles in the left with **a** 100 × magnification and **b** 700 × magnification

From the SEM image of the particles, it can be observed that they exhibit a relatively high tendency for agglomeration and an irregular morphology. The test specimens were fabricated using open silicone molds, as illustrated in the image below (Image 4).

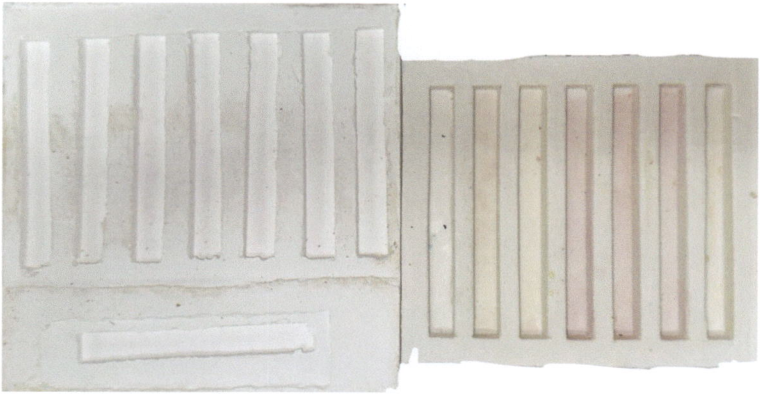

Image 4 Molds utilized in this work

The molds shown above correspond to the tensile test mold (left) and the flexural test mold (right), which were used for the mechanical tests performed in this study. Tensile tests were conducted in accordance with the ASTM D638-22 standard [15], while flexural tests followed the ASTM D790-17 standard [16]. The specimen dimensions were $120 \times 14.8 \times 3.2$ mm for the tensile test and $60 \times 12.5 \times 10$ mm for the flexural test.

The composites were prepared based on specific volume ratios, as presented in the table below. Fabrication was performed at ambient temperature, and the specimens were cured for approximately seven days prior to mechanical testing (Table 1).

Table 1 Formulations used for the fabrication of the composites

ID	Glass waste (%)	Epoxy resin (%)
EP	0	100
5VR	5	95
10VR	10	90
15VR	15	85
20VR	20	80

All mechanical tests were conducted using an Instron Model 5582 universal testing machine, located in the Mechanical Testing Laboratory of the Advanced Materials Laboratory (LAMAV) at UENF. A photograph of the Instron machine is shown below (Image 5).

Image 5 Universal machine INSTRON

Results and Discussion

The results are presented below, followed by a brief discussion. For the flexural test, the graph of the Young's modulus is shown below (Image 6).

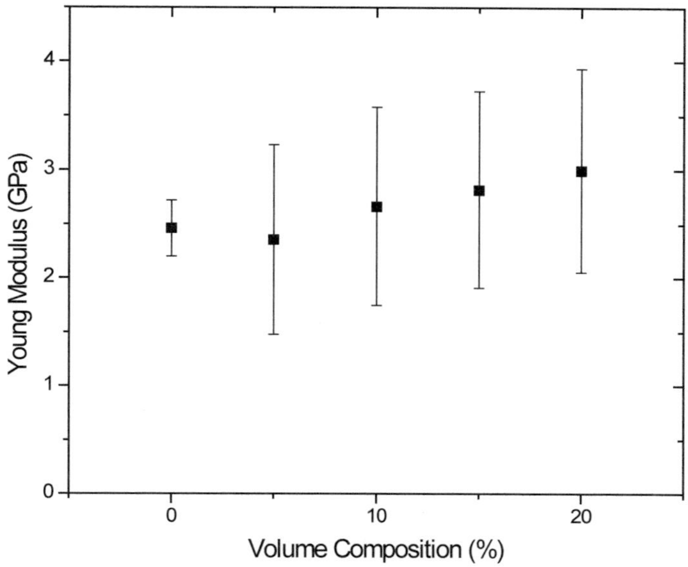

Image 6 Graph of the Young modulus in the 3 point bending test

The graph above indicates a trend of increasing Young's modulus with higher glass content. This behavior can be primarily attributed to the greater hardness of the glass particles, which, as previously discussed, directly affects both the hardness and the elastic modulus of the composite material [17].

Additionally, as the particle content increases, the mobility of the interlocking polymer chains is restricted, leading to greater brittleness and a reduced ability of the material to deform before failure [18]. The graph of maximum strength is presented below (Image 7).

As shown in the graph above, there is an apparent downward trend in the maximum stress sustained by the material. Although the observed deviations prevent a definitive conclusion regarding this reduction, the decrease in strength can be primarily attributed to the increased presence of defects and stress concentrators caused by higher particle content. These particles can impede the release of gases during the resin curing process [19].

Moreover, an inadequate interface between the particles and the matrix may lead to the formation of voids, which act as stress concentrators under tensile loading [20]. The graphs below present the Young's modulus and maximum stress values obtained from the tensile tests performed on the composites (Image 8).

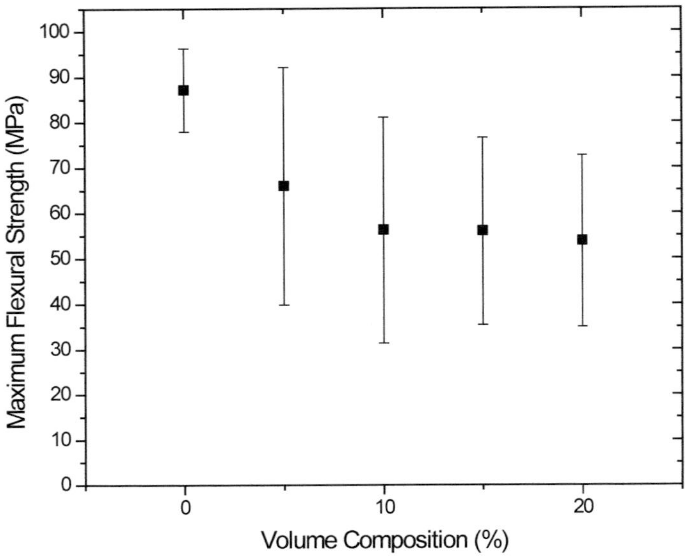

Image 7 Graph of the maximum flexural strength

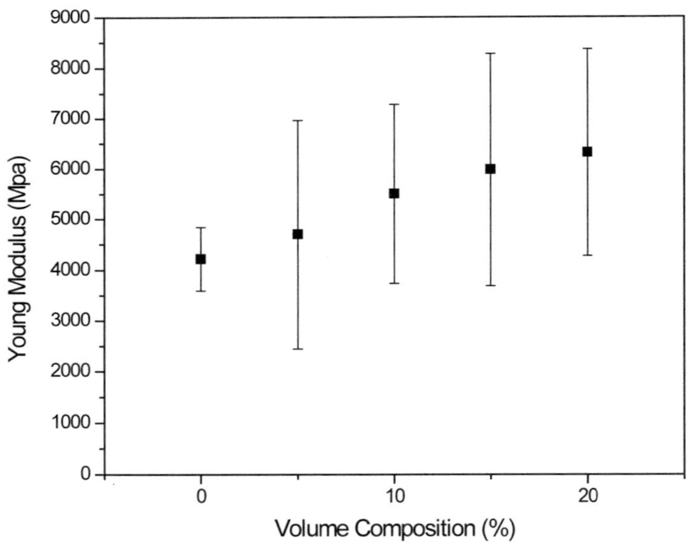

Image 8 Graph of the Young modulus in the tensile test

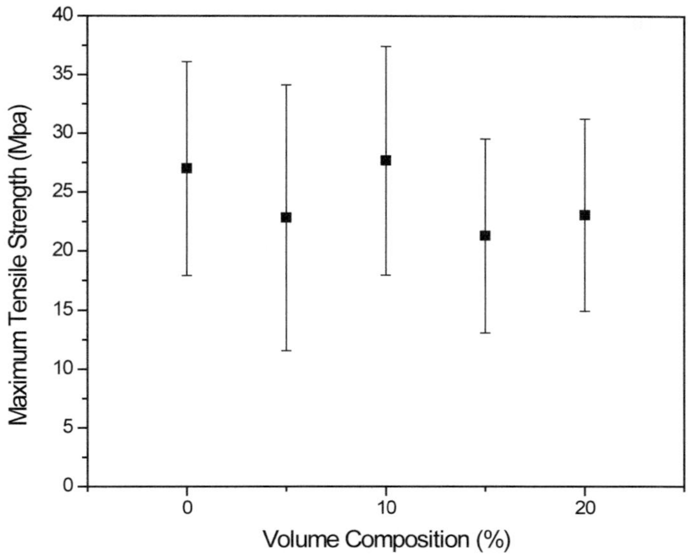

Image 9 Graph of the maximum tensile strength

As shown in the graph above, the modulus of elasticity exhibits an upward trend with increasing glass content, consistent with the results observed in the flexural test. This behavior can be attributed to the restriction of polymer chain mobility caused by the incorporation of rigid glass particles, which hinders molecular motion and consequently increases the modulus of elasticity. However, this effect also leads to a more brittle material [21]. The graph of maximum stress is presented below (Image 9).

The graph above shows a downward trend in maximum stress with increasing glass content, although considerable data dispersion is observed. While the composite containing 10% glass displayed the highest resistance, the large standard deviation across all samples indicates that the differences are not statistically significant; thus, all compositions can be considered to have comparable mechanical strength [22].

Tensile testing is a critical method for assessing the mechanical performance of composites, as it highlights the influence of stress concentrators under uniaxial loading. In this context, the incorporation of glass waste did not result in a statistically significant change in the tensile strength of the composites.

Conclusions

Based on the results, it can be concluded that the addition of glass waste did not lead to significant changes in the mechanical strength of the composites. A similar trend was observed for the modulus of elasticity, which, although showing a slight upward tendency, did not exhibit a statistically significant increase. Conversely, the maximum stress displayed a minor tendency to decrease. Nevertheless, the study successfully achieved its primary objective of developing a more environmentally friendly material.

The incorporation of glass waste as a dispersed phase directly reduced the amount of non-recyclable epoxy resin required. Additionally, since the dispersed phase consists of waste material, this substitution contributes to lowering the overall cost of the composite. As a result, the process produces a final material that is both more sustainable and more economical.

Acknowledgements We would like to thank CNPq for the financial support (Process No. 301634/2018-1) and FAPERJ for the support through the project (Ref. Proc. No. E-26/203.874/2022).

References

1. Görücü S, Kasirga ÜC, Akyol A, Weller A (2025) As an individual, I suppose you can't really do much: Environmental sustainability perceptions of machine learning practitioners. In: Proceedings of the 2025 ACM conference on fairness, accountability, and transparency, 1312–1324
2. Cook EM, Anderson EC, Childers DL (2025) Nature-based solutions for urban sustainability. Proc Natl Acad Sci U S A 122(29):e2315909122
3. Xie C, Zhang L, Wang Y (2025) Advances in the study of natural disasters induced by the "23.7" extreme rainfall event in North China. Nat Hazards Res
4. Baldassarre B, Carrara S (2025) Critical raw materials, circular economy, sustainable development: EU policy reflections for future research and innovation. Resour Conserv Recycl 215:108060
5. Paramasivam P, Alruqi M, Ağbulut Ü (2025) Waste to energy: enhancing biogas utilization in dual-fuel engines using machine learning based prognostic analysis. Fuel 381:133093
6. Xia R, Jia C, Garbatov Y (2025) Deterioration of marine offshore structures and subsea installations subjected to severely corrosive environment: a review. Mater Corros 76(6):758–775
7. Zhou T, Hu C (2025) A review on life-cycle cost analysis of anticorrosion coating systems relevance to its economic evaluation and performance insights for the structures in marine and offshore environment. Anti-Corros Methods Mater
8. Shojaei AR, Ramezanzadeh B, Mohammadloo HE (2025) Corrosion in the marine and offshore industry. In: Industrial corrosion: fundamentals, failure, analysis and prevention, pp 65–104
9. Tang Y, Li W, Zhang X, Zhao L (2025) Resorcinol–formaldehyde resin combining epoxy resin as composite coating with efficient anti-corrosion resistance and anti-biofouling capacity. Prog Org Coat 203:109177
10. Liu T, Zhao X, Wang J, Zhang Q (2025) Enhanced anti-corrosion and wear resistance of epoxy coatings through ploy (2-aminothiazole)-modified Mxene. Colloids Surf A Physicochem Eng Asp 137554

11. Li K, Zhang Y, Wang H, Liu W (2025) High-performance, recyclable, and degradable bio-based epoxy resins based on dynamic covalent imine bonds. J Appl Polym Sci 142(24):e57033
12. Chowaniec A, Sadowski Ł, Żak A (2020) The chemical and microstructural analysis of the adhesive properties of epoxy resin coatings modified using waste glass powder. Appl Surf Sci 504:144373
13. Chowaniec A, Ostrowski K (2018) Epoxy resin coatings modified with waste glass powder for sustainable construction. Tech Trans 115:99–109
14. Amaraibi RJ, Smith K, Johnson L, Davis M (2025) Experimental, economic, and life cycle carbon footprint assessment of low-cost adsorbents for siloxane removal from landfill gas. Environ Prog Sustain Energy 44(1):e14534
15. ASTM International (2022) Standard test method for tensile properties of plastics. ASTM D638–22. ASTM International, West Conshohocken, PA
16. ASTM International (2017) Standard test methods for flexural properties of unreinforced and reinforced plastics and electrical insulating materials. ASTM D790–17. ASTM International, West Conshohocken, PA
17. Kandas H, Ozdemir O (2024) An investigation of the mechanical characteristics of natural particle-reinforced glass and epoxy composites after immersion in acidic and basic aging solutions. Polym Compos 45(2):1288–1299
18. Xu P (2025) Topological regulation in polysilsesquioxanes for achieving super-hard and flexible membranes: insights from molecular simulation. Mater Horiz 12(5):1609–1618
19. Bai Z, Wang C, Li X (2025) Tensile fracture behavior and friction and wear properties of TiB2 particle reinforced Al-Si matrix composites. Ceram Int 51(1-B):991–1001
20. Fekiač JJ, Novakova A, Varga M (2025) Comprehensive review: optimization of epoxy composites, mechanical properties, & technological trends. Polymers (Basel) 17(3):271
21. Zhou C, Yang F, Wei Y (2025) In situ anchoring functional molecules to polymer chains through supramolecular interactions for a robust and self-healing multifunctional waterborne polyurethane. Small 21(10):2410933
22. Markides CF, Kourkoulis SK, Pasiou ED (2020) Enlightening the role of critical parameters for the determination of the tensile strength by means of the Circular Semi Ring test. Procedia Struct Integr 25:214–225

Evaluation of the Influence of Kraft Lignin on the Ultraviolet Degradation of Epoxy Composites with Epoxidized Soybean Oil

Paulo Roberto Correia Marcelino, Felipe Perissé Duarte Lopes, Hemanuelly Ferreira Breda Lan Oliveira, and Michel Picanço Oliveira

Abstract The increasing use of plastic materials has driven the search for eco-friendly alternatives, including plant-based composites aimed at reducing petrochemical-derived components. Epoxidized soybean oil (ESO), produced via epoxidation of soybean oil, is a renewable resource suitable for epoxy resin synthesis. However, conventional polymers and composites are susceptible to UV degradation, affecting their performance. In this study, kraft lignin was investigated as an additive in ESO-based epoxy composites (OSE20) for its UV-protective properties. Forty tensile specimens were fabricated with 20% ESO and varying lignin content (0%, 10%, 20%, and 30%). Half of the samples were exposed to UV radiation. Results showed that lignin effectively mitigated UV degradation, while samples without lignin exhibited greater damage. Despite this, mechanical strength decreased in all lignin-containing formulations. FTIR spectra showed no clear chemical changes. Thermal analysis revealed that lignin modified the degradation temperatures. Overall, lignin shows promise but requires further investigation to optimize mechanical performance.

Keyword Nanotechnology · Biomaterials · Composites

P. R. C. Marcelino · M. P. Oliveira (✉)
Forest and Wood Sciences Department, Federal University of Espírito Santo, Jeronimo Monteiro, Brazil
e-mail: michel.oliveira@ufes.br

F. P. D. Lopes
Materials Science and Engineering Department, State University of Northern Rio de Janeiro, Campos dos Goytacazes, Brazil
e-mail: perisse@uenf.br

H. F. B. L. Oliveira
Chemical Engineering Department, Federal University of Espírito Santo, Alegre, Brazil

H. A. C. Lopera et al. (eds.), *Natural Fibers and Biocomposites 2026*, The Minerals, Metals & Materials Series, https://doi.org/10.1007/978-3-032-13577-3_19

Introduction

The use of petroleum-derived materials has caused serious environmental impacts, prompting a significant demand for ecologically friendly materials to replace fossil-derived ones in the development of new products. This shift aims to reduce environmental impact, leading numerous studies to focus on utilizing natural, organic, and functional products as raw materials [1–3]. In this regard, certain plant-derived materials are being utilized in the development and synthesis of epoxy resins, with epoxidized soybean oil (ESO) derived from soybean oil being extensively studied. ESO is obtained through epoxidation, where its unsaturated double bonds are oxidized to form epoxy groups along the fatty acid chains. Its use shows promise in the development of natural-based epoxy resins [4–6].

Due to its abundance, high availability, and unique characteristics, lignin is another promising option among renewable resources. It is the second most abundant molecule in plant-derived materials, following only cellulose [7, 8]. In nature, lignin is a source of aromatic compounds. Its phenolic groups have the ability to scavenge free radicals, making it a material with interesting properties such as antibacterial and antioxidant actions. This confers lignin with the property of protection against ultraviolet (UV) rays [9].

UV light is a type of electromagnetic radiation present in sunlight, with wavelengths approximately ranging from 100 to 400 nm. It can initiate the formation of free radicals that lead to chain scission and cross-linking within the structure of polymeric materials [10]. Polymers do not have good resistance to UV radiation. When exposed to UV light, they present several disadvantages such as color loss or change, stains on their surface, and alterations in their internal properties. The absorbed radiation can break molecular bonds or create new chemical groups, thereby affecting mechanical characteristics such as increased brittleness, reduced impact resistance, and elasticity modulus [11, 12].

Most polymeric materials degrade when exposed to environmental conditions such as temperature, humidity, UV rays, and biological attack. In this context, lignin can be used as a UV protection agent. Its unsaturated functional groups, including carbon–carbon double bonds, aromatic rings, and conjugated carbonyl groups, absorb UV light, giving lignin its characteristic brown and black coloration while protecting the polymer from UV damage [13, 14].

Several studies have shown the potential of lignin as a UV protection agent. Goliszek et al. (2024) compared industrial and analytical kraft lignins as additives in UV protection coatings based on biological materials. They concluded that lignin enhanced the photoprotective properties of the coating and emphasized the valorization of lignin's potential for sustainable product applications [15]. Nasri et al. [13] evaluated the influence of UV irradiation on biocomposites with flax and pine fibers and stated that materials with higher lignin content showed less UV degradation [13]. Taher et al. [16] developed a nanocomposite film using lignin and epoxy resin with enhanced UV resistance performance. They described that the application of lignin resulted in increased UV blocking capacity and improved thermomechanical characteristics of the material [16].

Taking into consideration sustainability and the pursuit of green technologies, this study aims to replace part of the epoxy resin with epoxidized soybean oil and kraft lignin. The goal is to evaluate lignin's influence on the UV degradation of the material and position it as an innovative material to enhance the sustainability of polymeric materials. This approach promotes the use of renewable natural resources and reduces dependence on petroleum derivatives.

Methodology

In this study, composite materials were investigated, produced by polymerizing epoxy resin with 20% epoxidized soybean oil (ESO) and varying fractions of kraft lignin as additives. The selection of this specific percentage of ESO was based on findings from ongoing research within the current research group.

Materials

The epoxy resin used is coded as 2004 and specified as transparent and low viscosity. It was sourced from Redelease company, with the manufacturer recommending a 2:1 ratio of resin to hardener. The lignin was obtained from Klabin S/A, derived from bleached pine wood pulp. The epoxidized soybean oil was sourced from Gotalube Additives LTDA.

Production of Composites

For tensile testing, 40 specimens were produced and divided into four groups, with 10 samples each. Each group contained a different percentage of kraft lignin: 0, 10, 20, and 30%. Half of the specimens in each group were exposed to UV radiation. Table 1 illustrates the materials produced, their compositions, and UV exposure details.

The epoxy resin was mixed with epoxidized soybean oil, followed by kraft lignin dried for 24 h in an oven, and the hardener component. The molded samples were then allowed to rest for 24 h at room temperature and subsequently placed in a drying oven at 70 °C for 48 h to complete the curing cycle before characterization.

Table 1 Developed materials and their respective compositions

Treatment	Material code	Resin (%)	Epoxidized soybean oil (%)	Lignin (%)	UV exposure
OSE	OSE20	80	20	0	NO
LIGNIN	LIG010	70	20	10	
	LIG020	60	20	20	
	LIG030	50	20	30	
OSE	OSE20-UV	80	20	0	YES
LIGNIN	LIG010-UV	70	20	10	
	LIG020-UV	60	20	20	
	LIG030-UV	50	20	30	

UV Degradation Test

The composites were exposed to ultraviolet (UV) radiation for 20 days inside a box measuring 23 × 57 × 36 cm, using 10 JPN T8 model lamps with a power of 15 W and a wavelength of 253.7 nm. This analysis followed an adapted version of ASTM G151 (ASTM 2019) standard procedure.

Fourier Transform Infrared Spectroscopy (FTIR)

The FTIR analysis for lignin and the produced composites was conducted using a Bruker Tensor 27 spectrophotometer, located at the Analytical Central Laboratory (LEQ 3) at UFES Alegre campus. The analyses were performed in the frequency range of 4000 to 600 cm^{-1}, with a resolution of 4 cm^{-1}, employing attenuated total reflectance (ATR) technique.

Scanning Electron Microscopy

The evaluation of the fracture surface morphology of the composites was performed using scanning electron microscopy (SEM) with a JEOL JSM-IT200 instrument. For the analysis, samples were mounted on a metal base and coated with a thin layer of gold to enhance electrical conductivity in the material.

Mechanical Analysis

The tensile strength mechanical tests were conducted using an EMIC DL 100,000 universal testing machine, located in the mechanical testing laboratory at UFES Jerônimo Monteiro campus. The testing speed was set at 1 mm/min until sample rupture, following ASTM D638 (ASTM 2022) specifications.

Thermal Analysis

The thermal degradation behavior of the produced composites was evaluated using Thermogravimetric Analysis (TGA/DSC) technique. The analyses were conducted with a Setaram LabSys Evo Thermogravimetric Analyzer, located in the chemical engineering laboratory (LEQ 1) at UFES Alegre campus. Samples were heated from 25 to 800 °C at a heating rate of 10 °C/min under a nitrogen atmosphere.

Statistics

R Studio software was used to perform comparison of sample group distributions through analysis of variance (ANOVA) and Tukey's mean comparison test to assess significant differences among treatments.

Results and Discussion

Fourier Transform Infrared Spectroscopy (FTIR)

Figure 1 presents the FTIR spectra of the kraft lignin used in the development of the materials in this study. The spectra show a broad band around 3450 cm^{-1} corresponding to O–H stretching vibrations, while the band near 2930 cm^{-1} originates from C–H stretching vibrations. Bands around 1515 cm^{-1} are related to stretching vibrations of C–C and C = C bonds in the aromatic ring. Bands at 1260 cm^{-1} and 1030 cm^{-1} correspond to C–O–C ester group absorption and C–O stretching vibrations, respectively [17–19].

The spectra of the composites shown in Fig. 2 exhibited similar bands with changes in the intensity of some peaks. They included bands at 2922 cm^{-1} and 2852 cm^{-1} corresponding to C–H stretching in epoxy, 1738 cm^{-1} attributed to the carbonyl group in the triglyceride of epoxidized soybean oil (OSE), bands at 1610 cm^{-1} and 1509 cm^{-1} likely attributed to C = C stretching in aromatic compounds and stretching vibration in the aromatic ring, respectively. They also showed a band at 1232 cm^{-1}

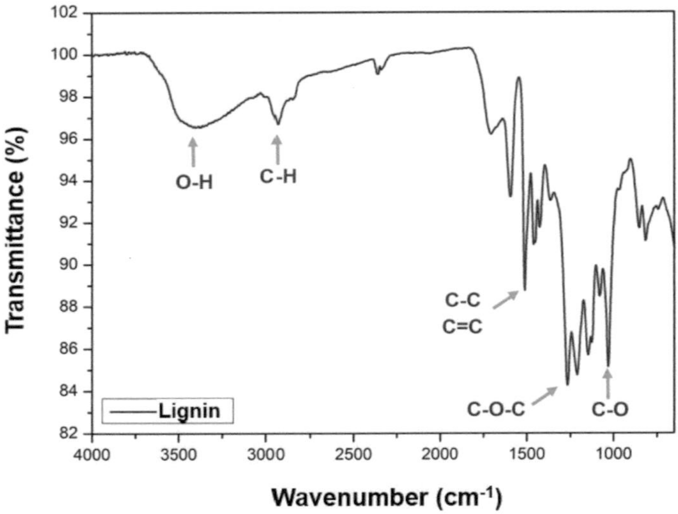

Fig. 1 FTIR spectra of lignin particles

related to ester linkage groups and/or oxirane rings, 1030 cm^{-1} attributed to CH2–OH group or C–O stretching vibration in lignin particles, and a band at 827 cm^{-1} originating from C–C vibration in the epoxy ring [19–21].

The difference between the spectra before and after UV exposure lies in the absorption intensity of some peaks, which showed a slight decrease after exposure. This decrease in absorption intensity at the 1505 cm^{-1} peak may be associated with possible opening of aromatic rings. Considering also the decrease at the 825 cm^{-1} peak, it suggests possible cleavage of C–C bonds during UV exposure. The reduction in intensity at the 2923 and 2853 cm^{-1} peaks indicates oxidation of susceptible hydrocarbons on the sample surface, as observed in other studies. [20, 22]. But these are not conclusive results, as this analysis is very sensitive to sample details such as color, particle size, and other characteristics, which can cause variations in band intensities.

Scanning Electron Microscopy

The images obtained by scanning electron microscopy of the fracture surface of the composites developed with different proportions of lignin before and after exposure to UV radiation for 480 h are presented in Figs. 3 and 4.

The main change indicated by the micrographs is the appearance of spots in various regions of the OSE20 matrix highlighted by rectangles in the images; these spots were not found in the other compositions. Polymer degradation can exhibit various characteristics, including surface yellowing and the appearance of spots,

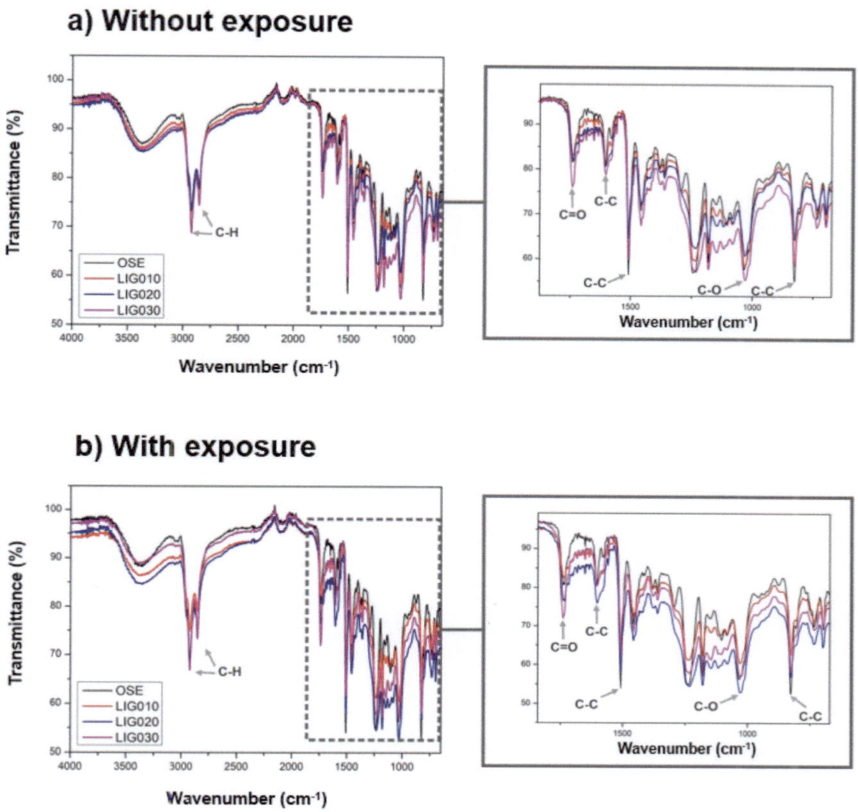

Fig. 2 FTIR spectra of the developed composites

which are correlated with structural changes occurring inside the polymer due to its degradation [11].

The appearance of these spots after 480 h of UV exposure only in the material without lignin is a morphological indicator that lignin indeed performed a protective function through the absorption of visible light, mainly due to its unsaturated functional groups, formed by C–C bonds, phenolic rings, and conjugated carbonyl groups, as observed in the FTIR, and supported by results reported by other authors [14].

The addition of lignin resulted in the appearance of a greater number of microcracks that can affect the mechanical strength of the composite. It is worth noting that this increase may be related to excessive agglomeration of lignin particles, which can form stress concentrators and negatively impact the mechanical characteristics of

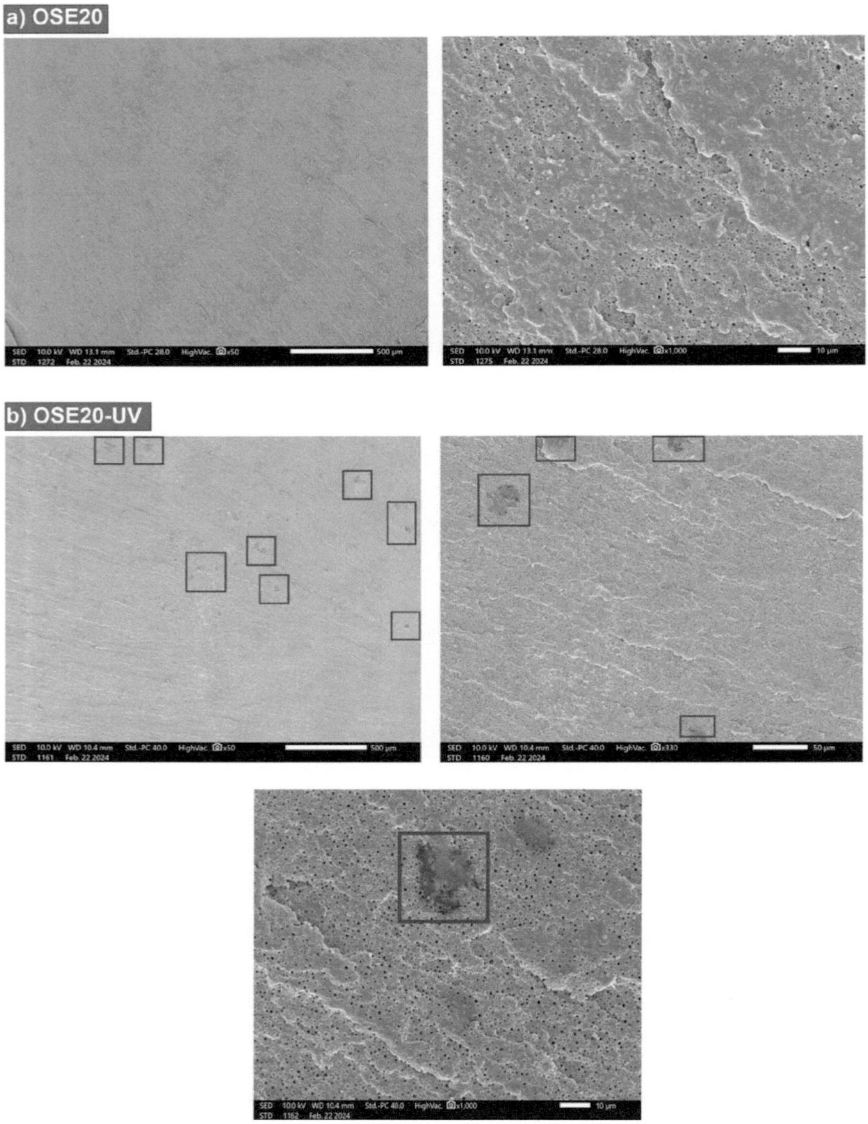

Fig. 3 Micrographs of the OSE20 matrix before and after UV exposure

the material, as commonly observed in particle-reinforced polymers [23]. However, particle agglomeration was not observed in the micrographs conducted in this study, and no significant differences were noted between the fracture surfaces of the composites with lignin before and after exposure to UV radiation.

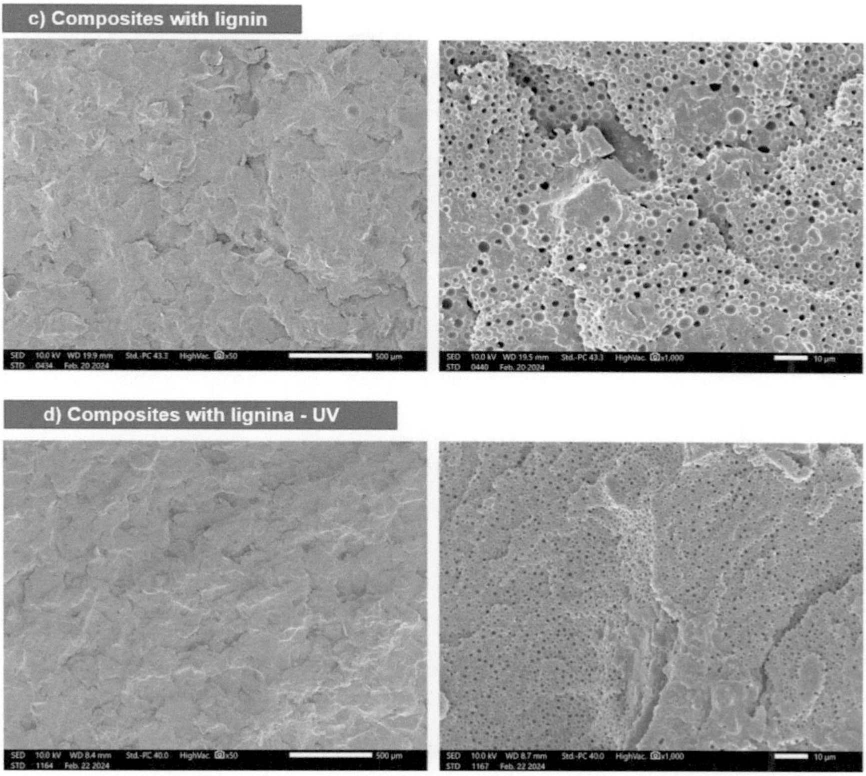

Fig. 4 Micrographs of the developed composites with lignin, before and after UV exposure

Mechanical Analysis

In Fig. 5, the results of the tensile test for the composites before and after UV exposure are presented. The letters represent groups that do not show statistically significant differences, according to the Tukey's mean comparison analysis.

It was observed that the incorporation of lignin had a significant negative impact on the mechanical strength of the developed composites. There was a substantial percentage loss of approximately 70% in tensile strength and 90% in toughness for the composites containing 30% lignin, before and after exposure to UV radiation. However, the incorporation of lignin did not significantly affect the modulus of elasticity, according to ANOVA and Tukey's mean comparison analyses. Similar results were observed by Taher et al. [16] in their study on nanocomposites based on nanolignin, where a decline in mechanical properties was also noted with the incorporation of lignocellulosic material, attributed to increased agglomeration of nanolignin load in the matrix [16].

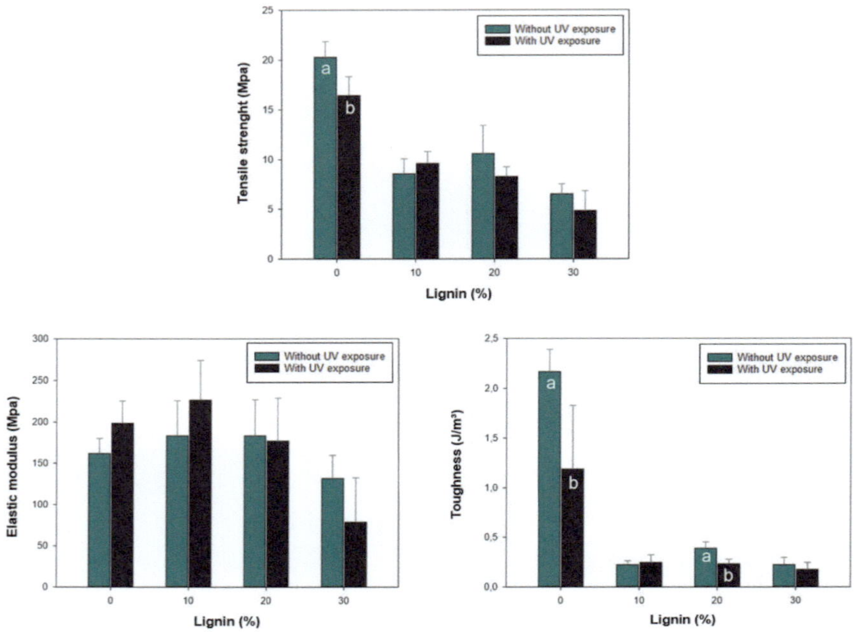

Fig. 5 Mechanical properties of the developed composites before and after UV exposure

The decrease in mechanical strength in all lignin-containing compositions may have been caused by particle agglomeration of lignin, although this phenomenon was not observed in the SEM micrographs. This agglomeration behavior can create stress concentration points, commonly observed in tensile tests of particle-reinforced polymers [23].

The UV protection provided by lignin was evidenced by comparing the maximum tensile strength of the composites before and after UV exposure. Composites without lignin showed a significant percentage loss of approximately 29% after exposure, whereas composites with lignin showed no significant variation in this property, according to statistical tests. Similar patterns were observed in the modulus of elasticity, where there was no significant variation in any of the compositions after exposure. Gualberto et al. [24] also reported similar patterns in the modulus of elasticity after UV exposure in epoxy matrix composites. However, in terms of toughness modulus, the composites with 0% and 20% lignin showed a significant decrease, with percentage losses close to 45% and 40%, respectively [24].

The decline in mechanical strength of materials without lignin may be related to the potential breakage of C–C bonds in the epoxy matrix during exposure, as evidenced by FTIR analysis. These bonds are primary in the epoxy matrix, and their decrease, along with an increase in C–O and C = O bonds, indicates an increase in material oxidation [25].

Thermal Analysis

The Fig. 6 shows the results obtained from the thermogravimetric analysis conducted on lignin for the developed composites.

Lignin begins its degradation around 180 °C, with its major degradation occurring in the range between 343 and 422 °C. According to the literature, lignin degradation occurs in three stages: the first stage, up to 180 °C, mainly corresponds to moisture loss; the second stage, spanning from 180 to 550 °C, represents the peak degradation of lignin, involving the breaking of chemical bonds within its structure and the release of various volatile compounds formed from the decomposition of its complex chains. The third stage occurs above 550 °C, involving the decomposition of more stable organic matter, resulting in an approximate residual mass of 47% [26, 27].

The OSE matrix exhibits a slight mass loss up to around 150 °C, which may be associated with moisture loss, and its main degradation range occurs between 310 and 470 °C in this study, related to depolymerization and degradation of its polymeric

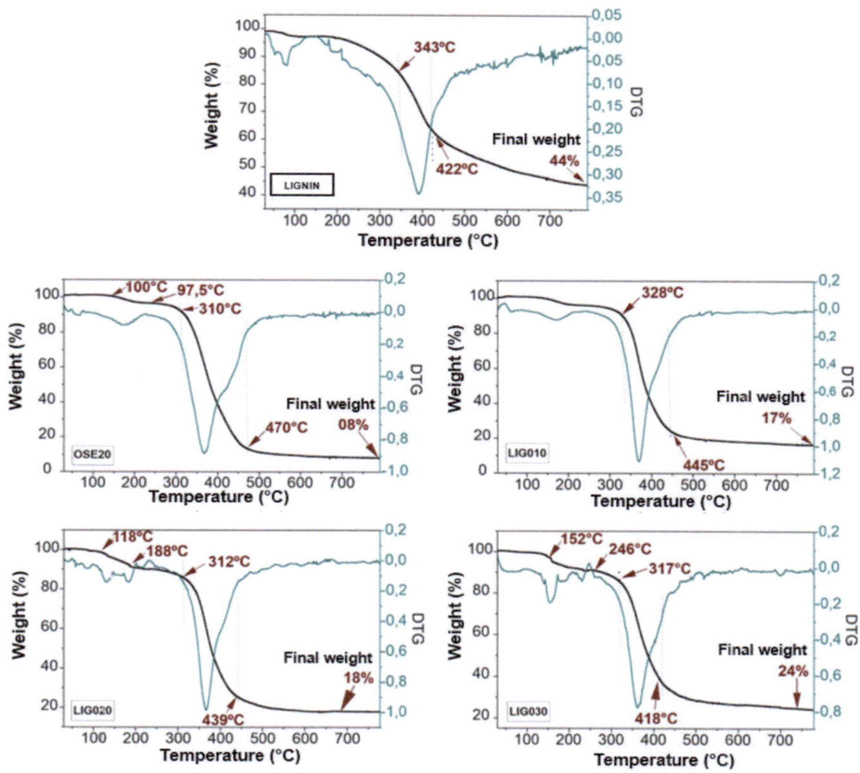

Fig. 6 TG, DTG, and DSC curves of lignin and the developed composites

chains [28]. The LIG010, 020, and 030 composites exhibited similar degradation behaviors, with some variations in the initial and final temperatures of the main degradation range. Lignin showed a significantly higher residual mass compared to OSE, which can be attributed to the higher carbon content present in lignin [15], Thus, the residual mass of the composites increased proportionally with the amount of lignin applied in their composition.

Conclusion

Lignin has proven effective in protecting against ultraviolet rays. This was justified by observing greater degradation of the OSE matrix in the absence of lignin compared to composites containing lignin, which showed no significant variations after UV exposure. The incorporation of lignin led to an increased presence of microcracks in the material, as observed in SEM micrographs. This behavior was attributed to possible lignin agglomeration within the matrix, indicating poor material dispersion, which explains the decrease in mechanical strength upon its incorporation.

FTIR spectral showed variations in the intensity of some bands; however, it is not appropriate to conclude that this caused chemical changes in the composites due to the sensitivity of the analysis to material characteristics. In thermal analyses, lignin influenced changes in the initial and final temperatures of the material's main degradation range. The results of lignin incorporation are promising, especially in UV protection, the focus of this study, but further studies and methodologies are needed to enhance lignin as a more effective reinforcement, particularly in enhancing mechanical properties of materials.

Acknowledgements This study was financed in part by the Espírito Santo Research and Innovation Support Foundation (FAPES), Brazil, through EDITAL FAPES N° 03/2021 - UNIVERSAL (grant numbers: 456/2021); EDITAL FAPES N° 28/2022 – UNIVERSAL (grant numbers: 932/2023) and National Council for Scientific and Technological Development – CNPq. Additionally, the Coordenação de Aperfeiçoamento de Pessoal de Nível Superior (CAPES) in Brazil provided funding under Finance Code 001.

References

1. Das O, Babu K, Shanmugam V et al (2022) Natural and industrial wastes for sustainable and renewable polymer composites. Renew Sustain Energy Rev 158:112054. https://doi.org/10.1016/j.rser.2021.112054
2. Solin K (2022) Nanocellulose interactions with protein and water in advanced sensing systems. In: Bioproducts ans biosystems
3. Bae S, Choi J-H, Ahn M et al (2023) Ethanol organosolv lignin as a substitute for commercial antioxidants, focusing on the structural properties and synergistic effect with myricetin. Food Chem 418:136009. https://doi.org/10.1016/j.foodchem.2023.136009

4. Sobhan A, Ahirekar V, Hoff M, Muthukumarappan K (2023) Derivation and characterization of epoxidized soybean oil and epoxy resin film produced using a three step-washing neutralization process. Ind Crops Prod 198:116675. https://doi.org/10.1016/j.indcrop.2023.116675
5. Zhen X, Cui X, Al-Haimi AANM et al (2024) Fully bio-based epoxy resins from lignin and epoxidized soybean oil: rigid-flexible, tunable properties and high lignin content. Int J Biol Macromol 254:127760. https://doi.org/10.1016/j.ijbiomac.2023.127760
6. Zhu F, Fu Q, Yu M et al (2023) Synthesis and curing properties of multifunctional castor oil-based epoxy resin. Polym Test 122:108017. https://doi.org/10.1016/j.polymertesting.2023.108017
7. Widsten P, Tamminen T, Liitiä T (2020) Natural sunscreens based on nanoparticles of modified Kraft Lignin (CatLignin). ACS Omega 5:13438–13446. https://doi.org/10.1021/acsomega.0c01742
8. Zhou Y, Qian Y, Wang J et al (2020) Bioinspired lignin-polydopamine nanocapsules with strong bioadhesion for long-acting and high-performance natural sunscreens. Biomacromol 21:3231–3241. https://doi.org/10.1021/acs.biomac.0c00696
9. Ariyanta HA, Santoso EB, Suryanegara L et al (2023) Recent progress on the development of lignin as future ingredient biobased cosmetics. Sustain Chem Pharm 32:100966. https://doi.org/10.1016/j.scp.2022.100966
10. Kuyyakanont A, Iwata M (2024) Study of different degradation effects in UV-sensitive polymers using xenon lamp and deuterium lamp to simulate UV irradiation in space environment. Nucl Instrum Methods Phys Res Sect B Beam Interact Mater Atoms 549:165267. https://doi.org/10.1016/j.nimb.2024.165267
11. Smoleń J, Olesik P, Nowacki B et al (2024) The influence of UV radiation on the properties of GFRP laminates in underwater conditions. Sci Rep 14:1–11. https://doi.org/10.1038/s41598-024-57999-8
12. Gao Q, Wang L, Luo H et al (2023) Photodegradation behavior and blocking strategy of waterborne polyurethane under UV and Xenon irradiation. Mater Today Commun 34:105212. https://doi.org/10.1016/j.mtcomm.2022.105212
13. Nasri K, Toubal L, Loranger É, Koffi D (2022) Influence of UV irradiation on mechanical properties and drop-weight impact performance of polypropylene biocomposites reinforced with short flax and pine fibers. Compos Part C Open Access 9:100296. https://doi.org/10.1016/j.jcomc.2022.100296
14. Paulsson M, Parkås J (2012) Review: light-induced yellowing of lignocellulosic pulps-mechanisms and preventive methods. BioResources 7:5995–6040
15. Goliszek M, Hosseinaei O, Matwijczuk A, Podkościelna B (2024) Comparing industrial and analytical kraft lignins as biobased UV-protective additives in coatings. Ind Crops Prod 214:118536. https://doi.org/10.1016/j.indcrop.2024.118536
16. Taher MA, Hasan KMF, Mahmud S et al (2024) Sustainable epoxy nanobiocomposites reinforced with lignin nanoparticles for enhanced UV resistance and mechanical properties. Compos Commun 48:101941. https://doi.org/10.1016/j.coco.2024.101941
17. Klapiszewski Ł, Jamrozik A, Strzemiecka B, et al (2017) Characteristics of multifunctional, Eco-friendly lignin-Al2O3 hybrid fillers and their influence on the properties of composites for abrasive tools. Molecules 22.https://doi.org/10.3390/molecules22111920
18. Jędrzejczak P, Puszka A, Kubiak A et al (2021) New lignin-based hybrid materials as functional additives for polymer biocomposites: from design to application. Int J Biol Macromol 190:624–635. https://doi.org/10.1016/j.ijbiomac.2021.09.025
19. Wang Y, Hou J, Huang Y, Fu Y (2022) Structure-controlled lignin complex for PLA composites with outstanding antibacterial, fluorescent and photothermal conversion properties. Int J Biol Macromol 194:1002–1009. https://doi.org/10.1016/j.ijbiomac.2021.11.159
20. Busiak R, Masek A, Węgier A, Rylski A (2022) Accelerated aging of epoxy biocomposites filled with cellulose. Materials (Basel) 15.https://doi.org/10.3390/ma15093256
21. Balaji A, Kannan S, Purushothaman R et al (2024) Banana fiber and particle-reinforced epoxy biocomposites: mechanical, water absorption, and thermal properties investigation. Biomass Convers Biorefinery 14:7835–7845. https://doi.org/10.1007/s13399-022-02829-y

22. Shi Z, Zou C, Zhou F, Zhao J (2022) Analysis of the mechanical properties and damage mechanism of carbon fiber/epoxy composites under UV aging. Materials (Basel) 15.https://doi.org/10.3390/ma15082919
23. Pereira BS, Barbosa R, Alves TS (2019) Evaluation of the morphology and mechanical properties of laminated composites based on epoxy, cork and glass microspheres. Rev Mater 24.https://doi.org/10.1590/S1517-707620190003.0754
24. Gualberto HR, dos Reis JML, de Andrade MC et al (2024) Effect of exposure time to UV radiation on mechanical properties of glass/epoxy composites. Appl Compos Mater 31:447–465. https://doi.org/10.1007/s10443-023-10182-0
25. Torres M, Burdin L, Rentería-Rodríguez AV, Franco-Urquiza EA (2023) Degradation of epoxy-particles composites exposed to UV and gamma radiation. Chem 5:559–570. https://doi.org/10.3390/chemistry5010040
26. Zhao J, Xiuwen W, Hu J et al (2014) Thermal degradation of softwood lignin and hardwood lignin by TG-FTIR and Py-GC/MS. Polym Degrad Stab 108:133–138. https://doi.org/10.1016/j.polymdegradstab.2014.06.006
27. Wądrzyk M, Janus R, Lewandowski M, Magdziarz A (2021) On mechanism of lignin decomposition–investigation using microscale techniques: Py-GC-MS, Py-FT-IR and TGA. Renew Energy 177:942–952. https://doi.org/10.1016/j.renene.2021.06.006
28. Da Silva TT, Mendonça Da Silveira PHP, Ribeiro MP et al (2021) Thermal and chemical characterization of kenaf fiber (Hibiscus cannabinus) reinforced epoxy matrix composites. Polymers (Basel) 13:1–15. https://doi.org/10.3390/polym13122016

Evaluation of the Properties of Polymer Composites with Amazônica Envira Cheirosa Fiber

M. V. F. Monteiro, D. S. Costa, A. P. S. Sousa, H. B. S. Lima, S. N. Monteiro, A. C. R. Silva, and V. S. Candido

Abstract This study presents an investigation of the performance of polyester matrix polymer composites reinforced with short fibers of envira cheirosa (Annona paludosa Aubl.), an Amazonian species still little explored in engineering. The research involved physical characterizations, flammability tests, and mechanical tensile tests, complemented by fractographic analyses using scanning electron microscopy (SEM). In this context, the aim was to understand the influence of short fiber addition on the structural, thermal, and mechanical properties of the composite. The results indicated that fiber incorporation increased porosity and water absorption, altered the burning behavior, and provided limited mechanical improvements, mainly at intermediate reinforcement concentrations. Microstructural analysis revealed interfacial adhesion failures, the presence of voids, and pull-out mechanisms, factors that explain the limitations of the observed mechanical properties. Thus, although envira cheirosa shows potential as reinforcement, the composite's efficiency directly depends on the quality of the fiber–matrix interface. Therefore, this study contributes to advancing knowledge on the application of Amazonian natural fibers in polymer materials, broadening perspectives for their sustainable use.

Keywords Polymer composites · Natural fibers · Amazônica fibers

M. V. F. Monteiro (✉) · D. S. Costa · A. C. R. Silva · V. S. Candido
Federal University of Pará–Ananindeua Campus/PA, Ananindeua, Brazil
e-mail: manoelviniciuz@gmail.com

D. S. Costa
e-mail: deibson@ufpa.br

A. C. R. Silva
e-mail: alissonrios@ufpa.br

V. S. Candido
e-mail: scarpini@ufpa.br

A. P. S. Sousa · H. B. S. Lima
Federal University of Pará–Belém Campus/PA, Belém, Brazil

S. N. Monteiro
Military Institute of Engineering/RJ, Rio de Janeiro, Brazil

H. A. C. Lopera et al. (eds.), *Natural Fibers and Biocomposites 2026*, The Minerals, Metals & Materials Series, https://doi.org/10.1007/978-3-032-13577-3_20

Introduction

The advanced materials sector faces the challenge of combining excellent technical performance with sustainability criteria. In this context, the search for alternatives to replace or complement synthetic reinforcements in polymer composites has driven significant growth in research involving plant fibers. Such fibers, being renewable, low-cost, lightweight, and presenting good specific properties, have been widely investigated as reinforcements for polymer matrices [11]. Furthermore, their use enhances the valorization of regional natural resources and contributes to the socioeconomic development of communities that depend on the extraction and processing of these raw materials.

In the Amazonian context, high biodiversity provides numerous species with potential for technological applications, although still little explored scientifically. Among them, envira cheirosa (Annona paludosa Aubl.) stands out, along with other natural fibers such as bamboo, tucum, piassava, and buriti, which have historical records of use in popular constructions and handicrafts, demonstrating structural resistance and durability [15].

The use of short fibers presents practical advantages related to processing, such as easier dispersion, lower cutting costs, and reduced alignment defects. However, the efficiency of the reinforcement is strongly conditioned by the fiber–matrix interfacial interaction, distribution homogeneity, and mitigation of internal defects, such as voids and agglomerates. These factors can act either positively or negatively on the final composite properties, depending on the mass fraction used and the processing conditions adopted.

Given this, the present study aimed to evaluate the behavior of polyester matrix polymer composites reinforced with short envira cheirosa fibers, considering physical properties, flame resistance, tensile mechanical performance, and microstructural aspects for sustainable industrial applications aligned with sustainability principles.

Materials and Methods

Materials

The polymer matrix adopted in this study was orthophthalic unsaturated polyester resin, catalyzed with methyl ethyl ketone peroxide (MEK-P) at a proportion of 1% by volume. The envira cheirosa fibers were obtained from a rural area in the municipality of Marapanim, located in the state of Pará, and subsequently cut into average lengths of approximately 15 mm, thus configuring the condition of short fibers. It is noteworthy that no prior chemical treatment was carried out, allowing in this context the evaluation of the intrinsic performance of the fiber in its natural form when incorporated into the polymer matrix. Figure 1 shows the envira cheirosa fiber at

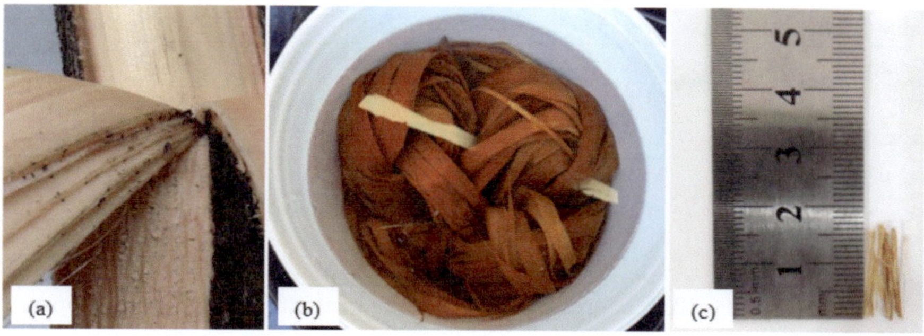

Fig. 1 Envira cheirosa fiber at different stages

three processing stages: fiber layer extraction (a), fiber immersion in water to delay degradation (b), and fiber cutting for addition into the mold (c).

Methods

The composites were produced by Hand Lay-Up method, using silicone molds, with fiber mass fractions of 1%, 3%, and 5% of short fibers. After curing, specimens were prepared for different tests. The physical tests included the determination of apparent density (AD), apparent porosity (AP), and water absorption (WA), conducted according to ASTM D792, D2734, and D570, respectively. The horizontal flammability test followed ASTM D635, providing the flame propagation rate in mm/min. The tensile mechanical performance was evaluated according to ASTM D638, using type I specimens. Finally, fractured samples were analyzed by scanning electron microscopy (SEM) in order to observe the predominant failure mechanisms.

Results and Discussion

Physical Tests

For the physical properties of the composites, data were obtained for Apparent Density (AD), given in g/cm^3, Apparent Porosity (AP), and Water Absorption (WA), expressed as percentages for the Pure Matrix (PM) and for the different compositions with short and long fibers. These results are presented in Table 1 with fiber contents of 1, 3, and 5%.

Based on the results presented in Table 1, it can be observed that the composites with short fibers exhibit similar AD values, a behavior that can be explained by the fact that the envira fiber has a density similar to that of the resin, which,

Table 1 Physical test data for composites with short fibers

Compositions (%)	ASM (%)	AP (%)	AA (%)
MP – 0%	1,154 ± 0,02	4,865 ± 1,74	4,226 ± 1,58
FC – 1%	1,163 ± 0,01	3,356 ± 0,45	2,878 ± 0,39
FC – 3%	1,168 ± 0,02	4,173 ± 1,72	3,559 ± 1,55
FC – 5%	1,176 ± 0,02	4,541 ± 1,59	3,890 ± 1,41

as described by Zuo [21], is predicted by the rule of mixtures. Regarding apparent porosity, a reduction is noted, as well as in water absorption, for all compositions containing fibers. This reduction can be attributed to the good homogenization of the fiber–matrix mixture, resulting in fewer pores, bubbles, and material defects. Furthermore, it is important to highlight the direct correlation between water absorption and apparent porosity, as both are related to the number of voids in the material. The incorporation of fibers into the matrix directly impacts these properties [7, 16].

Flammability Test

Table 2 presents the burning rates of the composites with envira cheirosa fiber, obtained through the flammability test conducted according to ASTM D635.

In terms of the absolute values presented in Table 2, it can be observed that the composites reinforced with short envira cheirosa fibers exhibited a slight increase in the flame propagation rate compared to the pure matrix. However, it is important to highlight that the incorporation of fibers into the composites not only resulted in a reduction in the amount of resin used, thereby lowering manufacturing costs, but also provided environmental benefits.

When comparing the flame propagation rate results of these composites with the requirements established by Resolution No. 498 of 2014 from the National Traffic Council (CONTRAN) and ASTM D 635, it was verified that both the pure matrix and the composites containing envira fibers performed within the safety parameters. The flame propagation rate remained below the maximum limits, shown in Table 3, allowed by both standards, demonstrating compliance with the criteria required for applications involving fire resistance. Table 3 presents a comparison of the best flammability test result with values reported in the literature.

Table 2 Results of the flammability test for the composites

Compositions (%)	Burning rate (mm/min)
MP – 0%	24,09 ± 1,26
FC – 1%	25,51 ± 0,62
FC – 3%	28,77 ± 1,55
FC – 5%	28,32 ± 1,81

Table 3 Comparison of the flammability test from this study with the literature

Compositions (%)	Burning rate (mm/min)	Source
5% F. de Envira Cheirosa	25,51 ± 0,62	This work
5% F. Bamboo	33,40 ± 0,25	[12]
20% F. Pineapple	31,40 ± 1,40	[18]
ASTM D635	40,0	–
CONTRAN	100,0	–

The composite manufactured with envira cheirosa fiber stands out, primarily because no flame retardants were used in its production process, unlike the composite made with pineapple fiber, which also uses a polyester matrix. According to Rocha, flame retardants are chemical substances employed to delay or, when possible, prevent the spread of flames in materials. However, these substances can be harmful to human health, as fine particles may be inhaled from the products in which they are applied, in addition to causing environmental damage.

Thus, it can be observed that the incorporation of envira cheirosa fibers into the matrix was effective, since all obtained values remained below the limits established by ASTM D 635, being classified as HB (Horizontal Burning), which indicates burning rates below 40 mm/min. Furthermore, according to the National Traffic Council (CONTRAN) resolution, it is recommended that interior coatings in the automotive industry present a maximum flame propagation rate of 100 mm/min. In this sense, the composites developed in this study comply with these requirements, as their flame propagation rates remained below the limits established by the standards.

Tensile Test

For the mechanical properties of the composites, the average tensile strength values, evaluated according to ASTM D638, for the pure matrix (PM) and the different compositions with short fibers are presented in Table 4.

According to the data presented in Table 5, it was observed that the composites with short fibers showed a progressive increase in strength as the fiber content in the matrix increased. However, the three evaluated proportions (1%, 3%, and 5%) resulted in a decrease in strength compared to the pure matrix. This indicates that the short envira cheirosa fiber predominantly acted as a filler material in the composite.

Table 4 Tensile test results of composites with short fibers

Compositions (%)	Tensile strength (MPa)	Elongation (%)
MP – 0%	34,83 ± 1,24	15,71 ± 5,12
FC – 1%	22,60 ± 1,95	5,63 ± 0,42
FC – 3%	23,10 ± 1,53	5,04 ± 1,62
FC – 5%	27,89 ± 1,11	8, 68 ± 2,46

Table 5 Comparison of tensile results with the literature

Compositions (%)	Tensile strength (MPa)	Source
5% F. de Envira Cheirosa	27,89 ± 1,11	This work
15% F. de Coco	15,34 ± 0,79	Tasgın et al. [17]
10% F. de bananeira + Casca de arroz	17,42 ± 1,09	[1]

This behavior can be attributed to poor fiber distribution in the matrix, low interfacial adhesion, or the presence of voids in the material. Nevertheless, from a sustainability perspective, the use of plant fibers in composites is advantageous, as they are renewable and biodegradable resources, contributing to the environmental sustainability of the products in which they are applied, in addition to being a more economical alternative compared to synthetic reinforcements [13, 20]. Table 5 presents a comparison of the results with literature data on polymer composites with the same matrix containing different proportions of short fibers.

The tensile strength obtained for the composite with 5% short envira cheirosa fiber showed superior performance compared to composites reported in the literature, such as the epoxy reinforced with 15% coconut fiber reported by Taşgın et al. [17] and the hybrid containing 10% banana fiber and rice husk described by [1]. This difference indicates that envira cheirosa exhibits higher reinforcement efficiency even without chemical treatment, possibly due to its rougher fibrous morphology, which promotes better interfacial mechanical adhesion and more effective stress transfer between the fiber and the matrix. Furthermore, the homogeneous distribution of fibers within the polyester matrix and the low fiber mass fraction contributed to minimizing internal defects, resulting in superior properties compared to other plant-based composites.

Fractographic Analysis of Tensile Failure

Through the analysis of the fracture surface of a material using SEM, it was possible to observe and characterize the presence of imperfections such as voids, bubbles, and cracks, and correlate them with the failure mechanisms and their possible causes. Figure 2 shows the microscopy of the fracture surface of composites with short envira cheirosa fibers after the tensile test.

On the fracture surface of the composite with 1% short fibers, Fig. 2a, small cracks (orange arrows) and voids (blue arrows) can be observed, which may indicate material weaknesses. These defects suggest that the composite has stress concentration points, which reduce mechanical strength and facilitate crack propagation under mechanical loading [8].

In the composite with 3% short fibers, Fig. 2b, the fracture is more pronounced (green arrow), with evidence of microcracks and failures following a preferred direction (orange circle). This pattern indicates that the material underwent brittle fracture, characterized by rapid propagation of failures along the fracture plane. Although

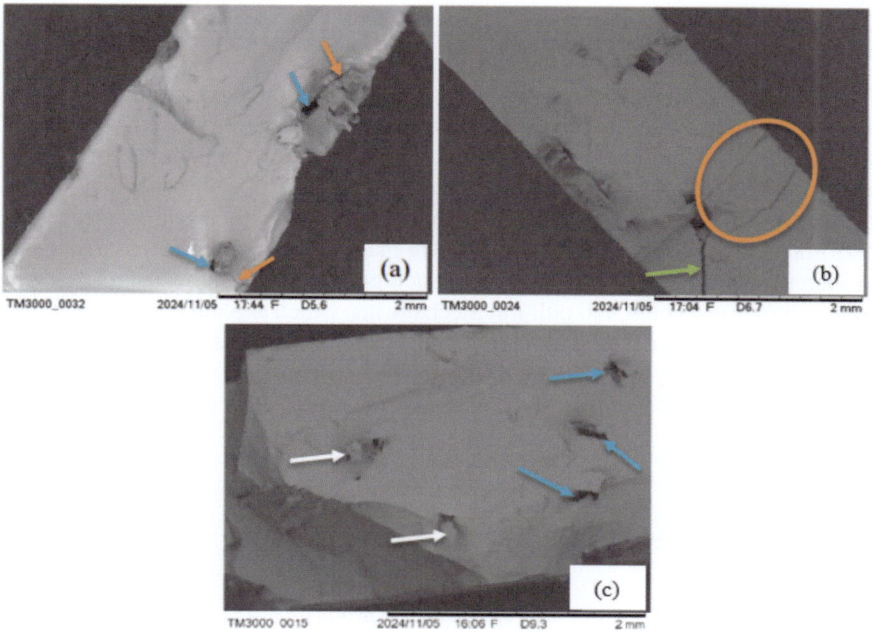

Fig. 2 Fractographic analysis of the composite with short fibers at 1% (**a**), 3% (**b**), and 5% (**c**)

cracks are a natural failure mechanism in loaded materials, the presence of imperfections such as microcracks and directional failures negatively affects the composite's mechanical strength. These imperfections may be associated with the manufacturing method, especially the distribution of fibers in the matrix during the fabrication process [8].

In the fracture of the composite with 5% fibers, Fig. 2c, the presence of voids (blue arrows) and pulled-out fibers (white arrows) can be observed, possibly indicative of the "pull-out" phenomenon. However, fractured fibers (white circles) are also present, suggesting that these fibers, although partially, exhibited good fiber/matrix interface, increasing tensile strength as the fiber mass fraction in the matrix increased. When the crack propagates in the direction normal to the fiber, the fibers separate from the matrix, resulting in complete composite failure. This process occurs under tensile loading when the applied stress or strain exceeds the maximum capacity supported by the fibers. In other words, the fibers fracture upon reaching their maximum strength or elongation capacity [14].

Conclusions

Polyester matrix composites reinforced with short envira cheirosa fibers exhibited apparent density similar to that of the pure matrix, along with reduced porosity and water absorption, indicating good system homogenization. In flammability tests, a slight increase in flame propagation rate was observed, but all values remained below regulatory limits, ensuring safety for applications. Conversely, tensile mechanical performance was lower than that of the pure matrix, although it showed a tendency to increase with higher fiber fractions. Fractographic analysis revealed interfacial adhesion failures, voids, and pull-out mechanisms, which explain the observed limitations. Thus, envira cheirosa shows potential as reinforcement in polymer composites, particularly for applications that prioritize sustainability and cost reduction, provided that improvements in the manufacturing process and fiber/matrix interface are implemented.

References

1. Adediran AA et al (2024) Mechanical properties of hybrid banana fibre, rice husk and eggshell reinforced epoxy composite. Res Eng Struct Mater (RESM) 10(2):458–466
2. ASTM D2734–16 (2016) Standard test methods for void content of reinforced plastics. ASTM International, West Conshohocken
3. ASTM D570–98 (2018) Standard test method for water absorption of plastics. ASTM International, West Conshohocken
4. ASTM D635–18 (2018) Standard test method for rate of burning and/or extent and time of burning of plastics in a horizontal position. ASTM International, West Conshohocken
5. ASTM D638–14 (2014) Standard test method for tensile properties of plastics. ASTM International, West Conshohocken
6. ASTM D792–20 (2020) Standard test methods for density and specific gravity (relative density) of plastics by displacement. ASTM International, West Conshohocken
7. Barbosa GP (2021) Avaliação de compósitos poliméricos reforçados com fibras vegetais: propriedades físicas e mecânicas. Universidade Federal de Pernambuco, Recife, Dissertação de Mestrado
8. Bezerra FJ (2022) Caracterização microestrutural e mecânica de compósitos reforçados com fibras naturais. Universidade Federal do Ceará, Fortaleza, Dissertação de Mestrado
9. Boaretto V (2023) Propriedades e aplicações de resinas poliéster insaturadas em compósitos. Rev Mater 28(3):1–12. https://doi.org/10.1590/1980-5373-MR-2022-0101
10. Conselho Nacional de Trânsito (CONTRAN) (2014) Resolução nº 498, de 29 de julho de 2014. Dispõe sobre requisitos aplicáveis aos materiais de revestimento interno dos habitáculos de veículos automotores. Disponível em: Plataforma Gov.br/Denatran. Acessado em 01 de set. 2025
11. Khalid MY (2021) Natural fiber reinforced polymer composites: sustainable materials for the future. Compos Part C Open Access 3:100100
12. Lima HBS (2022) Avaliação da resistência à chamas de compósitos poliéster isoftálico insaturado com resíduo de lama vermelha e fibra de bambu. Trabalho de Conclusão de Curso (Bacharel em Engenharia de Materiais)–Universidade Federal do Pará, Ananindeua
13. Maiti S et al (2022) Sustainable fiber-reinforced composites: a review. Adv Sustain Syst 6(1):2200258

14. Mulenga TK, Rangappa SM, Siengchin S (2025) Natural fiber composites under impact loading: a critical review. eXPRESS Polym Lett 19(5):470–493
15. Potiguara RCV et al (1987) Plantas fibrosas: levantamento botânico na microrregião do Salgado. MPEG-PA, Belém
16. Santos RM (2020) Influência da porosidade em compósitos poliméricos reforçados com fibras lignocelulósicas. Universidade Estadual de Campinas, Campinas, Dissertação de Mestrado
17. Tasgin Y, Demircan G, Kandemir S, Acikgoz A (2024) Mechanical, wear and thermal properties of natural fiber-reinforced epoxy composite: cotton, sisal, coir and wool fibers. J Mater Sci 59:10844–10857
18. Ribeiro LM et al (2023) Flamabilidade e retardância de chama do compósito: poliéster insaturado reforçado com fibra de abacaxi (palf). Holos 1:115–126
19. Rocha IG (2015) Influência do retardante de chamas nas propriedades mecânicas e flamabilidade em compósitos poliméricos de epóxi/fibra de curauá. (Dissertação de Mestrado)-Universidade Federal do Rio Grande do Norte
20. Silva R (2022) Aplicações de fibras lignocelulósicas na química de polímeros e em compósitos. Quim Nova 32(3):661–671
21. Zuo Q, Wang C, Pei X, Lin L, Li Y, Sun W (2023) Analysis and prediction of tensile properties based on rule of mixtures model for multi-scale ramie plain woven fabric reinforced composite. Compos Struct 317:116785

Flexural and Hardness Behavior of Ramie Fiber-Reinforced Composites with and Without Sweat Absorption

Luís Fernando Fortunato de Freitas, Victor Paes Dias Gonçalves, Michel Picanço Oliveira, Noam Tonini Simonassi, Ana Paula da Fonseca Piedade, and Felipe Perissé Duarte Lopes

Abstract This work investigates the flexural and hardness behavior of epoxy composites reinforced with ramie fibers, with and without sweat absorption. Composites containing 0%, 40%, and 60% fiber volume fractions (vol%) were produced and subjected to mechanical tests and morphological analysis. Results show that ramie fiber addition improves flexural strength, stiffness, and surface hardness compared to pure resin, with the 40% fraction exhibiting the best balance between reinforcement and structural integrity. However, artificial sweat absorption significantly reduces flexural properties, mainly due to fiber swelling and matrix–fiber interface degradation, as evidenced by SEM analysis. In contrast, hardness was less affected by sweat exposure, indicating greater surface stability. The main objective of this study is to evaluate the influence of artificial sweat absorption on the flexural and hardness behavior of epoxy composites reinforced with different ramie fiber contents. These findings highlight the potential of ramie fiber composites for sustainable applications, while also emphasizing the importance of moisture control to ensure long-term performance.

Keywords Ramie fiber · Epoxy resin · Natural composite · Flexural strength · Hardness · Sweat absorption · Biodegradable material

L. F. F. de Freitas (✉) · V. P. D. Gonçalves · N. T. Simonassi · F. P. D. Lopes
Advanced Materials Laboratory—LAMAV, State University of the Northern Rio de Janeiro—UENF, Av. Alberto Lamego, 2000, Campos dos Goytacazes, RJ 28013-602, Brazil
e-mail: fortunatodefreitas2@gmail.com

M. P. Oliveira
Department of Forestry and Wood Sciences, Federal University of Espírito Santo, Avenida Governador Lindemberg, Jeronimo Monteiro, ES, Brazil

A. P. da Fonseca Piedade
Department of Mechanical Engineering, University of Coimbra, CEMMPRE, Rua Luís Reis Santos, Coimbra, Portugal

H. A. C. Lopera et al. (eds.), *Natural Fibers and Biocomposites 2026*, The Minerals, Metals & Materials Series, https://doi.org/10.1007/978-3-032-13577-3_21

Introduction

The possibility of using advanced technologies in the development of new materials has opened up new avenues for integrated planning and interdisciplinary projects, in line with the Industry 4.0 concept [1, 2]. Increasingly, technology aims to facilitate, accelerate, and improve sustainable workflow, promoting energy efficiency and waste reduction [3, 4]. Few areas of materials engineering and additive manufacturing have evolved as rapidly in recent years as composite materials, keeping pace with globalization and offering more innovative features and environmentally friendly solutions [5, 6].

Polymer composites are increasingly subject to stringent requirements in terms of mechanical properties, especially in scenarios where efficiency and durability are essential [7, 8]. Natural fibers and their composites have been increasingly used due to their excellent properties, particularly with regard to low cost and sustainability aspects [9, 10]. There are many types of natural fibers that reinforce composites, with almost every country having its own natural fibers [11, 12]. In the field of plant biomechanics, the main focus is to evaluate mechanical characteristics with the aim of using this information to create new structures or derivative products in appropriate engineering projects [12, 13].

With the growing demand for materials capable of withstanding extreme conditions and complex applications, these composites must offer not only strength, but also flexibility, toughness, and lightness [14, 15]. The ability of polymer composites to meet these requirements, offering superior performance and reliability, is crucial for their application in advanced sectors such as aerospace, automotive, and medical [15, 16]. Therefore, this study aims to evaluate the influence of artificial sweat absorption on the flexural and hardness behavior of epoxy composites reinforced with different ramie fiber volume fractions (0%, 40%, and 60%), providing a better understanding of how moisture exposure affects their mechanical performance and interfacial integrity.

Materials and Methodology

The ramie fibers used in this study were purchased commercially from a distribution center in the state of Pará, Brazil, as illustrated in Fig. 1.

The test specimens were reinforced with 0%, 40%, and 60% [fiber volume fractions (vol%)] fiber content and were produced at room temperature and cured for 24 h. Rectangular plates of these composites, measuring 149.28 mm × 127.93 mm × 4.67 mm, were molded in a closed steel mold. The ramie fibers were positioned in the mold according to the precise calculation of the required amount. Next, the epoxy resin was mixed with the catalyst at a PHR of 16%, initiating the process of cross-linking the polymer chains and curing the polymer, which will become rigid [17]. During the curing process, a load of one ton was applied to the mold cover to

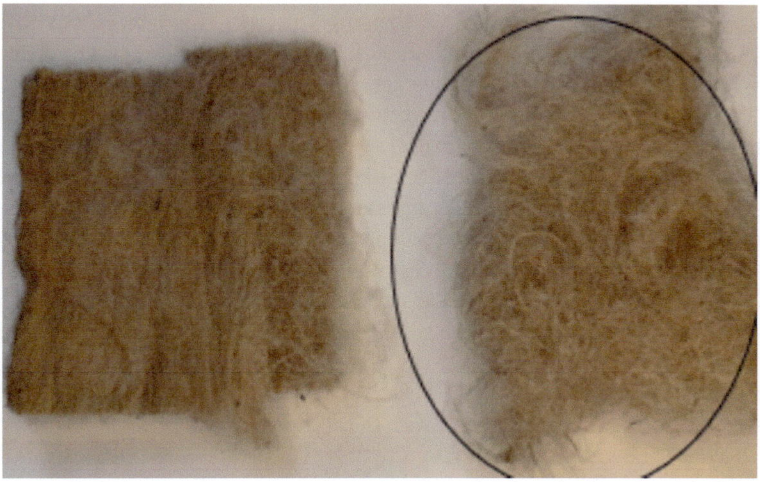

Fig. 1 Clean fiber

ensure adequate impregnation of the resin between the fibers [18]. After curing for one day, each plate was cut along the direction of the fibers, resulting in five test specimens for each formulation without immersion in artificial sweat and three test specimens for immersion in artificial sweat. For the flexural test, the ties were 80 mm long.

The test specimens were immersed in an artificial sweat solution composed of NaCl, lactic acid ($C_3H_6O_3$), and urea (CH_4N_2O) in an orbital shaker (100 rpm) at 37 °C, using 20 mL of the solution per cycle. After exposure, there was no rupture of the test specimens, but the formation of a viscous layer was observed on the surfaces, more evident in the samples with 60% fibers and in the lateral cut regions, where the fibers were more exposed.

The flexural tests were performed on a universal testing machine (Shimadzu AG–X 1 kN) at a speed of 2 mm/min, following ISO 14125 and ISO 178 standards, at room temperature (≈25 °C). The hardness tests (Shore D and Shore A) were performed on a portable durometer, with five readings per sample. The fracture surfaces were analyzed by scanning electron microscopy (SEM) using a TESCAN MIRA 4 microscope operating at 10 keV after gold sputter coating.

Results

The analysis of the results obtained shows the significant impact of the addition of ramie fibers and the absorption of simulated sweat on the mechanical behavior of the composites. Regarding deformation, the reinforced composites (CP-40% and CP-60%) presented values higher than those of pure resin, suggesting greater energy absorption capacity and adaptation to bending stresses. This behavior is favorable for applications that require strength associated with a certain degree of ductility. However, the absorption of artificial sweat resulted in a decrease in deformation, indicating that the presence of moisture makes the material more fragile and less capable of dissipating energy. Finally, the flexural stress analysis shows that composites with ramie fibers perform better than pure resin, especially CP-40%, which achieved the highest strength values. This result suggests that the 40% fraction may represent an adequate balance between reinforcement and structural integrity. Nevertheless, exposure to artificial sweat reduced the stress of all composites, confirming that moisture consistently compromises mechanical strength, especially due to the deterioration of the fiber–matrix interface and the weakening of the reinforced structure (Figs. 2, 3, 4).

In relation to Young's modulus, it can be observed that the presence of fibers increases the stiffness of the material, with the composite with 60% fiber volume fraction (CP-60%) being stiffer than that with 40% (CP-40%). However, after exposure to artificial sweat (CP-40%-ASA and CP-60%-ASA), there is a significant reduction in stiffness, reflecting the hydrophilic nature of ramie fiber and the consequent degradation of the fiber–matrix interface, which compromises stress transfer.

The analysis of the hardness results shows that the addition of ramie fibers, in both volumetric fractions (40% and 60%), increases surface resistance in relation to pure resin, confirming the positive effect of reinforcement. Regarding artificial sweat absorption, it is observed that the variation is insignificant in Shore A hardness, while in Shore D there is only a slight reduction, more evident in the composite with 60% fibers. These results indicate that hardness is less sensitive to the presence of moisture when compared to flexural strength, reflecting its predominantly surface character and lesser dependence on the integrity of the fiber–matrix interface.

Scanning electron microscopy (SEM) analysis provides fundamental information about fracture mechanisms and the quality of the fiber–matrix interface in ramie-reinforced composites, especially after exposure to artificial sweat. The micrographs shown in Figs. 5, 6, 7, 8 and 9 clearly illustrate the influence of fiber content and wet conditioning on fracture morphology. In pure resin (Fig. 5), a relatively homogeneous and smooth surface is observed, with crack propagation marks and the presence of bubbles, revealing the brittle nature of the polymer matrix. The absence of reinforcement results in cleaner and more direct fractures, with no evidence of additional energy dissipation mechanisms.

With the addition of 40% ramie fibers (Fig. 6), the fracture surface becomes rougher and more heterogeneous, with fibers distributed throughout the matrix and pull-out occurring, indicating that fiber–matrix adhesion is not ideal. Voids and porosity are visible, suggesting limitations in the impregnation process, which can reduce the efficiency of the reinforcement.

After exposure to artificial sweat (Fig. 7), these deficiencies become more pronounced: there is increased fiber–matrix detachment, a greater number of voids, and fibers with signs of superficial swelling. This scenario confirms that moisture absorption weakens the interface, favoring the propagation of cracks and justifying the loss of mechanical strength observed in the flexural tests.

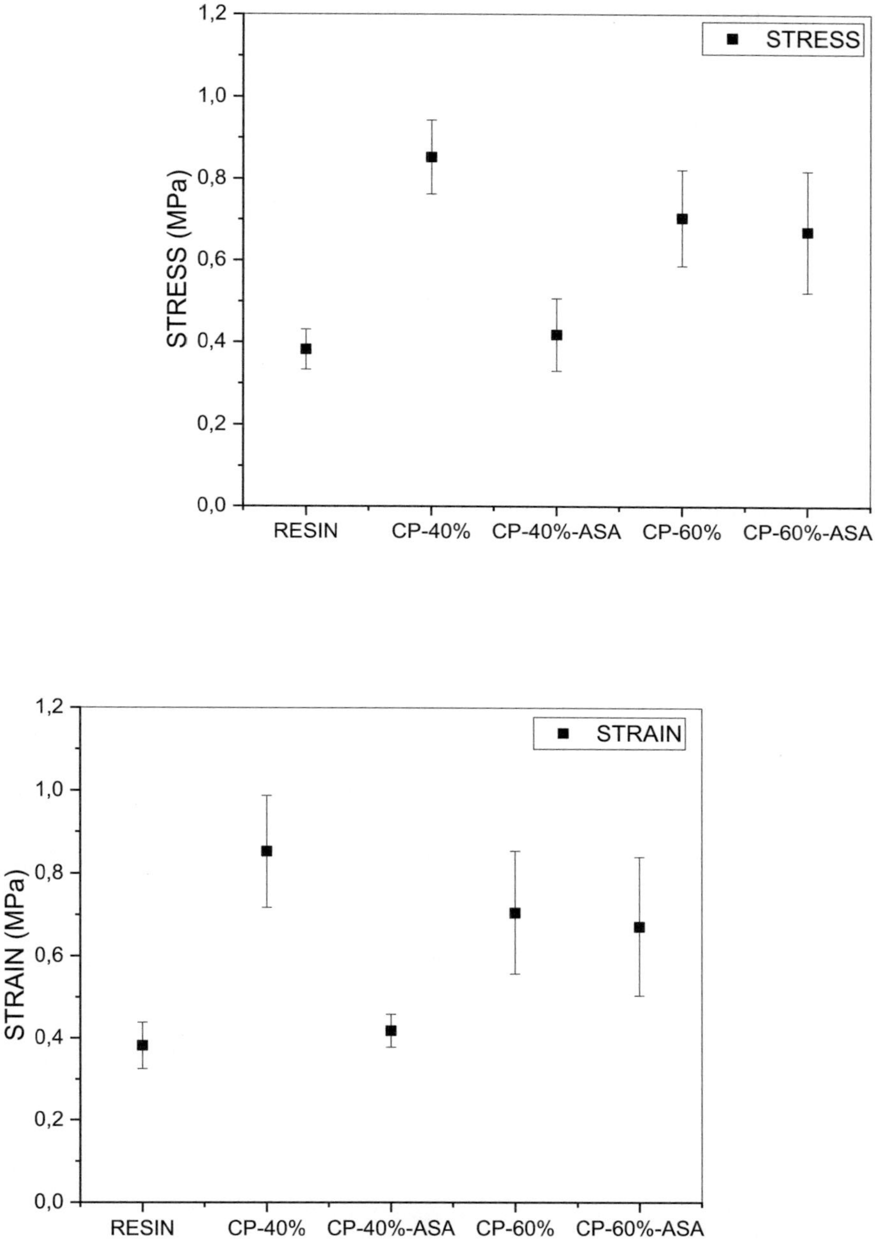

Fig. 2 **a** Flexural stress of ramie and pure resin composites, with and without simulated sweat absorption. **b** Flexural strain of ramie and pure resin composites, with and without simulated sweat absorption

Fig. 3 Young's modulus of ramie and pure resin composites under bending, with and without simulated sweat absorption

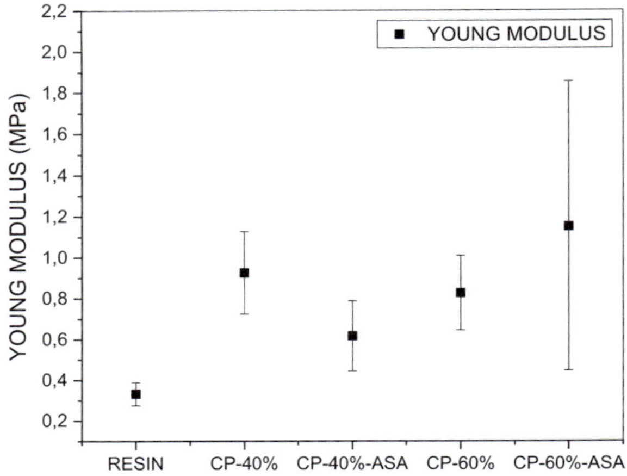

Fig. 4 Shore A (HA) and Shore D (HD) hardness values of ramie and pure resin composites, with and without simulated sweat absorption

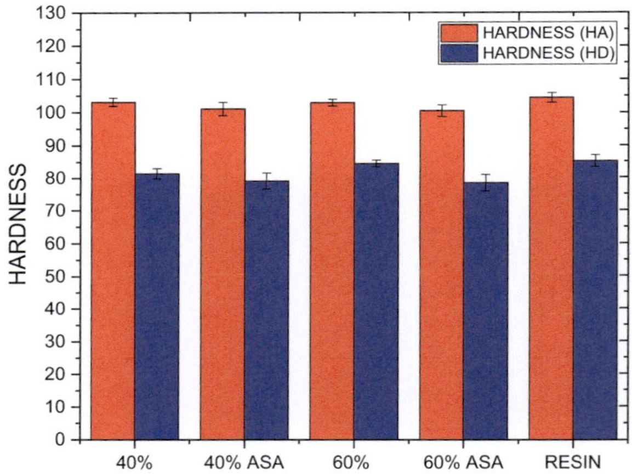

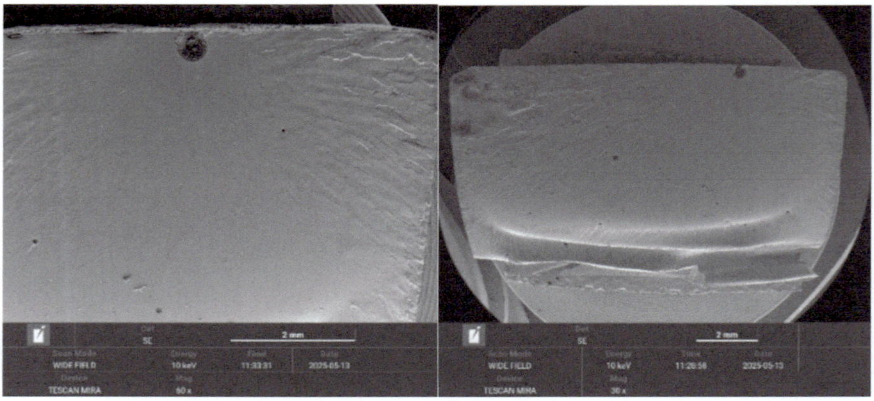

Fig. 5 **a** and **b** SEM micrographs of the pure resin test specimen

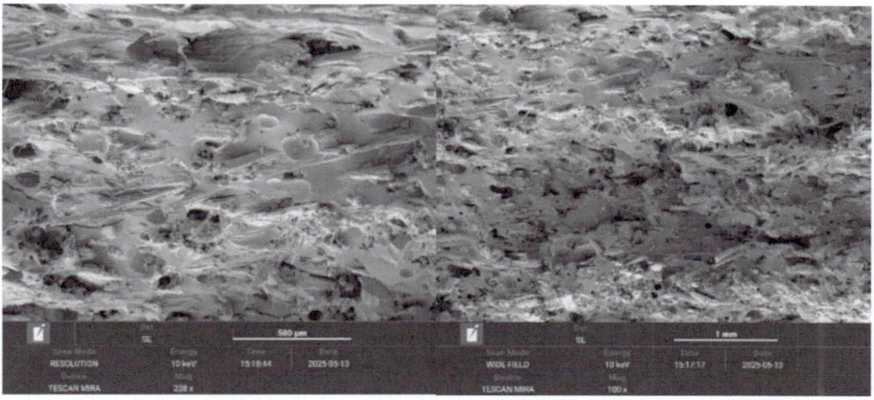

Fig. 6 **a** and **b** SEM micrographs of the resin composite test specimen with 40% ramie reinforcement

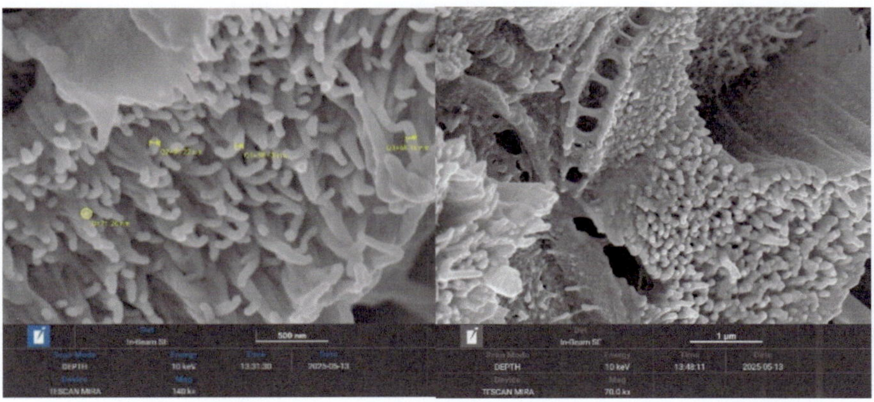

Fig. 7 **a** and **b** SEM micrographs of the resin composite test specimen with 40% ramie reinforcement after immersion in artificial sweat

In the composite with 60% fibers (Fig. 8), the fracture has a higher density of exposed fibers and a more compact network, but still with adhesion failures in some regions. Although the high presence of fibers can improve stiffness, it also increases the likelihood of agglomerations and regions of poor impregnation, creating critical points for crack nucleation.

After immersion in artificial sweat (Fig. 9), the degradative effects become even more evident, with intensified pull-out and a large number of voids around the fibers. The swelling of the fibers, combined with the weakening of the matrix, generates internal stresses that accelerate the rupture of the interface, resulting in a significant loss of structural integrity.

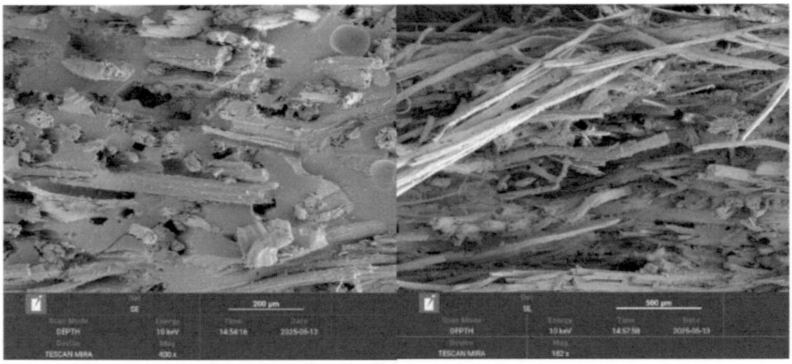

Fig. 8 **a** and **b** SEM micrographs of the resin composite test specimen with 60% ramie reinforcement

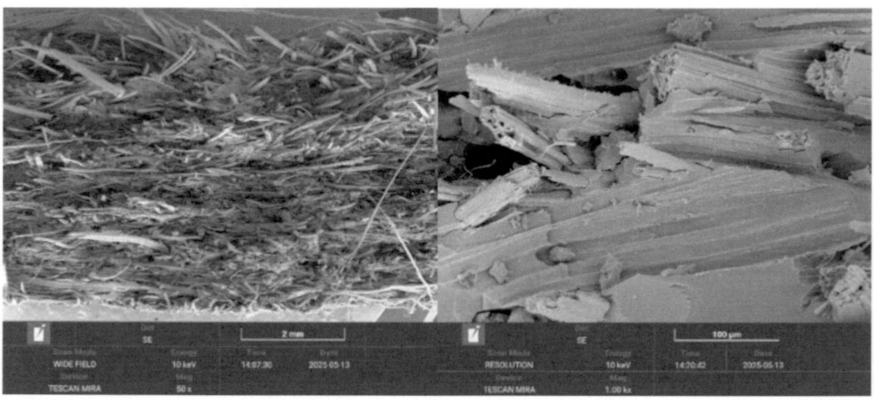

Fig. 9 **a** and **b** SEM micrographs of the resin composite test specimen with 60% ramie reinforcement after immersion in artificial sweat

The SEM observations provide a clear correlation between the morphological damage and the mechanical degradation observed in the flexural and hardness results. Together, the SEM results confirm the mechanisms responsible for the reduction in mechanical properties after sweat absorption. While the addition of fibers promotes greater roughness and additional fracture mechanisms, moisture compromises adhesion at the interface, enhancing the propagation of failures. Thus, the micrographs directly corroborate the results obtained in the flexural and hardness tests, showing that water degradation is a determining factor for the durability of composites reinforced with natural fibers.

Conclusion

The study demonstrated that the incorporation of ramie fibers into epoxy resin enhances flexural performance and hardness, reinforcing the potential of this natural fiber as a sustainable alternative in composite development. Among the evaluated formulations, the 40% fiber volume fraction (vol%) showed the most favorable mechanical response, balancing reinforcement and ductility. However, exposure to artificial sweat caused a notable reduction in flexural strength and stiffness, mainly due to interface weakening and fiber swelling, as confirmed by SEM analysis. Hardness, in turn, was less sensitive to sweat absorption, reinforcing its surface-dependent nature. Overall, the results emphasize both the advantages and limitations of ramie-reinforced composites, underlining the need for protective treatments or barrier strategies to improve their durability in humid or sweat-exposed environments.

References

1. Gonçalves VPD, Vieira CMF, Colorado HA (2024) Mechanical behavior of tension of multi-polymers through fused deposition modeling. In: TMS 2024 153rd annual meeting & exhibition supplemental proceedings. The minerals, metals & materials series. Springer, Cham. https://doi.org/10.1007/978-3-031-50349-8_12
2. Sánchez-García E, Martínez-Falcó J, Marco-Lajara B, Manresa-Marhuenda E (2024) Revolutionizing the circular economy through new technologies: a new era of sustainable progress. Environ Technol Innov 33:103509
3. Gonçalves VPD, Vieira CMF, Colorado Lopera HA (2024) The production and materials of mouthguards: conventional vs additive manufacturing—a systematic review. Heliyon 10. http://creativecommons.org/licenses/by-nc/4.0/
4. Javaid M, Haleem A, Singh RP, Suman R, Gonzalez ES (2022) Understanding the adoption of industry 4.0 technologies in improving environmental sustainability. Sustain Oper Comput 3:203–217
5. Asyraf MRM, Ng LF, Khoo PS, Yahya MY, Hassan SA, Madenci E, Kha T (2024) Lignocellulosic abaca fiber-reinforced thermoplastic composites as future sustainable structural materials: a bibliometric analysis and literature review. Cellulose 31:5419–5459. https://doi.org/10.1007/s10570-024-05921-w
6. Martinelli FRB, Ribeiro FRC, Marvila MT, Monteiro SN, Filho FCG, Azevedo ARG (2023) A review of the use of coconut fiber in cement composites. Polymers 15:1309. https://doi.org/10.3390/polym15051309
7. Bin Rashid A, Haque M, Mohaimenul Islam SM, Rafi Uddin Labib KM (2024) Nanotechnology-enhanced fiber-reinforced polymer composites: recent advancements on processing techniques and applications. Heliyon 10(2)
8. Noureddine M (2019) Study of composite-based natural fibers and renewable polymers, using bacteria to ameliorate the fiber/matrix interface. J Compos Mater 53:455–461
9. Pereira AC, Assis Fd, Garcia Filho FdC, et al (2019) Evaluation of the projectile's loss of energy in polyester composite reinforced with fique fiber and fabric. Mater Res 22:e20190146
10. Geitmann A, Niklas K, Speck T (2019) Plant biomechanics in the 21st century. J Exp Botany 70:3435–3438
11. Restrepo Carmona A, Colorado HL (2020) Luffa fibers as promising reinforcement for polymer composites: mechanical characterization of NaOH treated and untreated dumbbell test-pieces with Weibull statistics. J Compos Mater 0(0):1–15. https://doi.org/10.1177/0021998320976800

12. Santos TF, Santos CM, Aquino MS, Suyambulingam I, Hussein EK, Verma A, Rangappa SM, Siengchin S, Nascimento JHO (2024) Towards sustainable and ecofriendly polymer composite materials from bast fibers: a systematic review. Eng Res Express 6(1):012501. https://doi.org/10.1088/2631-8695/ad2640
13. Kwiatkowska D, Burian A (2022) Plant Biomechanics-a natural transition from molecular to organ scale. Int J Mol Sci 23(14):7575. https://doi.org/10.3390/ijms23147575.PMID:35886922;PMCID:PMC9320334
14. Parveez B, Kittur MI, Badruddin IA, Kamangar S, Hussien M, Umarfarooq MA (2022) Scientific advancements in composite materials for aircraft applications: a review. Polymers 14:5007. https://doi.org/10.3390/polym14225007
15. Khan F, Hossain N, Mim JJ, Rahman SM, Iqbal MJ, Billah M, Chowdhury MA (2024) Advances of composite materials in automobile applications—a review. J Eng Res. In press, corrected proof. Available 27 Feb 2024
16. Nahmias HLM, Vieira RM, Goncalves VPD, Freire BB, Chilvarquer I, Feghali FJ, Azevedo MMA (2022) The use of perioguide in surgery to correct gingival smile. Res Soc Dev 10(2)
17. Simonasssi NT, Monteiro SN (2019) Análise De Weibull Da Resistência À Tração De Compósitos De Alto Desempenho De Matriz Epóxi Reforçada Com Fibras Naturais De Curauá. In: 74° Congresso Anual da ABM, São Paulo, pp 946–955. ISSN: 2594-5327. https://doi.org/10.5151/2594-5327-33314
18. Monteiro SN, Aquino RCMP, Lopes FPD, de Carvalho EA, d'Almeida JRM (2006) Compósitos De Matriz Poliéster Reforçados Com Fibras Contínuas E Alinhadas De Curauá. In: 61 Congresso Anual da ABM, Rio de Janeiro, pp 550–558. ISSN: 2594-5327. https://doi.org/10.5151/2594-5327-0065

Formulation and Characterization of Starch-Based Bioplastics with Sargassum–Sawdust Additives via Factorial Design

Aline Hernandez-Garcia, Darinel Ortiz-Montaño, Alejandra Dominguez-Morales, Milton Elias-Espinosa, Yazmin Aguirre-Macias, Pedro Perdigon-Lagunes, and Iyari Nava-Téllez

Abstract This study reports the development of starch-based biopolymer films using Sargassum spp. (1.3–10% w/w) and sawdust (8.5–13.6% w/w) as fillers, following a 2-factor, 4-level factorial design. The optimal formulation, with a 35:16 mass ratio of sawdust to sargassum, exhibited a density of 1.3 g/mL, a 3.5% moisture content, a 19.4% solubility, a 27.3% water absorption, and a 40% biodegradability. Mechanical testing indicated improved flexibility with moderate tensile strength. TGA confirmed thermal stability, while X-ray diffraction (XRD) analysis revealed semi-crystalline features attributed to cellulose and alginate structures. SEM–EDS revealed a heterogeneous but cohesive morphology. These findings suggest strong compatibility among the components and highlight the potential of this biopolymer for use in packaging and agricultural applications, contributing to biomass valorization and sustainable materials development within a circular economy framework.

Keywords Biomaterials · Characterization · Sustainability · Bioplastic

A. Hernandez-Garcia (✉) · D. Ortiz-Montaño · A. Dominguez-Morales · Y. Aguirre-Macias
Anáhuac University, Av. Universidad Anáhuac, State of Mexico, Huixquilucan de Degollado, Mexico
e-mail: aline.hernandez@anahuac.mx

M. Elias-Espinosa
Tecnologico de Monterrey School of Engineering and Sciences, Monterrey, Mexico

P. Perdigon-Lagunes · I. Nava-Téllez
Tecnologico de Monterrey, Mexico City, Mexico

H. A. C. Lopera et al. (eds.), *Natural Fibers and Biocomposites 2026*, The Minerals, Metals & Materials Series, https://doi.org/10.1007/978-3-032-13577-3_22

Introduction

The increase in the production and consumption of petroleum-derived plastics has led to a global environmental crisis, particularly due to their persistence in terrestrial and marine ecosystems [1, 11]. It is estimated that single-use plastics, such as those used in food packaging, account for up to 36% of the total manufactured, making them one of the leading causes of environmental pollution [9]. Faced with this problem, biopolymers have emerged as a sustainable alternative, as they are biodegradable and can be obtained from renewable resources or biomass waste [2, 4, 8].

Among the different sources explored for obtaining biopolymers, brown algae of the genus Sargassum have gained relevance due to their high availability and their content of alginates and polysaccharides with film-forming potential [5, 9]. This resource is especially attractive in regions such as the Caribbean and the Atlantic, where massive arrivals of Sargassum spp. generate negative impacts on tourism, fishing, and public health, but at the same time represent an abundant biomass susceptible to valorization [7, 10].

On the other hand, lignocellulosic waste, such as sawdust, has also been widely studied as a reinforcement in polymeric matrices due to its low cost, abundance, and ability to modify the structural and mechanical properties of the materials [3, 6]. The combination of both resources—Sargassum spp. and sawdust—offers an opportunity for the development of hybrid bioplastics that leverage both marine polysaccharides and the lignocellulosic fraction, thereby contributing to circular economy models.

Previous studies have reported the production of bioplastics from sodium alginate, which exhibits good oxygen barrier properties and high biodegradability [9, 11], as well as the feasibility of designing integrated processes to produce biopolymers from organic waste [8]. However, most of this research has focused on the use of individual raw materials, leaving an area of opportunity in combining different biomasses to optimize the performance of the final material.

In this context, this work aims to develop and characterize starch-based biopolymer films reinforced with Sargassum spp. and sawdust, evaluating their physicochemical, structural, thermal, mechanical, and biodegradation properties. This aims to establish a preliminary technological route for the joint valorization of these biomasses in the production of biodegradable materials.

Materials and Methods

Preparation of Biopolymer Films

Starch-based biopolymer films were developed using Sargassum spp. (1.3, 3.5, 5.7, and 10% w/w) and sawdust (8.5, 9.8, 11.1, and 13.6% w/w) as reinforcing agents, following a two-factor factorial design with four levels. The optimal formulation corresponded to a mass ratio of sawdust to sargassum of 35:16, using as a matrix a mixture consisting of 10.6% corn starch, 6% sodium bicarbonate, 17.7% 10% v/v acetic acid solution, and the remainder distilled water.

The mixtures were manually homogenized and then heated on a hotplate at 200 °C until a change in density was observed. The resulting masses were poured into specific molds according to the test to be performed.

Physical–Chemical Characterization

Density

Cubic specimens measuring approximately 1 × 1 × 0.5 cm were prepared. Each sample was weighed, and the calculated volume was used to determine the material's apparent density.

Moisture Content

Test specimens of equal size were weighed and dried in an induction oven at 80 °C for 48 h. They were then stored in a desiccator until room temperature was reached and reweighed to calculate the moisture content.

Solubility in Water

The test tubes were weighed and immersed in 5 mL of distilled water for five hours. After removing excess surface moisture, they were allowed to dry at room temperature for 24 h before being weighed again. The loss in mass was used to calculate the percentage solubility.

Water Absorption

Test tubes were immersed in distilled water for five h. After removing excess moisture, the samples were immediately weighed and then dried at room temperature to determine the weight gain attributable to water absorption, differentiating it from the possible dissolution of components.

Biodegradation in Soil

The test tubes were buried in compostable soil at room temperature for 20 days. At the end of the period, the samples were recovered, cleaned, and weighed to calculate the percentage of mass loss attributable to biodegradation.

Structural and Thermal Characterization

X-Ray Diffraction (XRD)

2 × 2 × 0.5 cm specimens were molded and analyzed in an X-ray diffractometer (Cu-Kα), with a scanning range 2 θ of 5–40°, at a speed of 0.5°/s (0.02°/min). The samples with the best physical properties were selected for evaluation.

Table 1 Code to identify the samples

Sargassum/sawdust [%w,w]	8.5	9.8	11.1	13.6
1.3	S1A1	S1A2	S1A3	S1A4
3.5	S2A1	S2A2	S2A3	S2A4
5.7	S3A1	S3A2	S3A3	S3A4
10	S4A1	S4A2	S4A3	S4A4

Thermogravimetric Analysis (TGA)

A sample of 1 g was used. Analyses were performed within a range of 30–900 °C, with a heating rate of 10 °C/min under a nitrogen atmosphere (50 mL/min).

Scanning Electron Microscopy and Microanalysis (SEM–EDS)

Specimens measuring 0.5 × 0.5 × 0.5 cm were mounted on double-sided conductive carbon tape and coated. Analyses were performed using a scanning electron microscope (Zeiss) under variable pressure, equipped with an EDS detector (Bruker) for elemental microanalysis.

A specific code identified each sample, the description of which is presented in Table 1.

Results

This section presents the results obtained from the preparation and characterization of biopolymer films developed with Sargassum spp. and sawdust as reinforcement materials. The analyses were organized according to the physical, chemical, structural, and degradation properties evaluated, to establish the influence of the different formulations on material performance. The data are presented in comparative tables and figures that allow for identifying trends, contrasting variations between treatments, and determining the formulation with the most significant potential for applications in biodegradable packaging or filters.

The response surfaces obtained from the factorial experimental design (DOE) are presented below, allowing for the visualization of the influence of Sargassum spp. and sawdust concentrations on the physical properties of the biopolymer films. Figure 1 shows the variation in density as a function of both factors; Fig. 2 illustrates the behavior of the solubility percentage; Fig. 3 represents the humidity percentage; and Fig. 4 reflects the water absorption percentage. These graphical representations facilitate the identification of trends and the determination of optimal formulation conditions to maximize material performance.

Figure 5 presents the X-ray diffractograms (XRD) corresponding to the selected samples with the best physicochemical properties, as determined by the DOE. The characteristic peaks of starch, polylactic acid (PLA), and microcrystalline cellulose

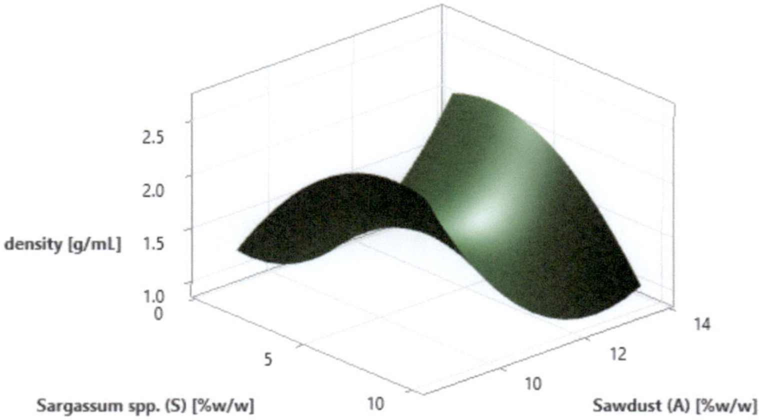

Fig. 1 Response surface for density according to the proposed DOE

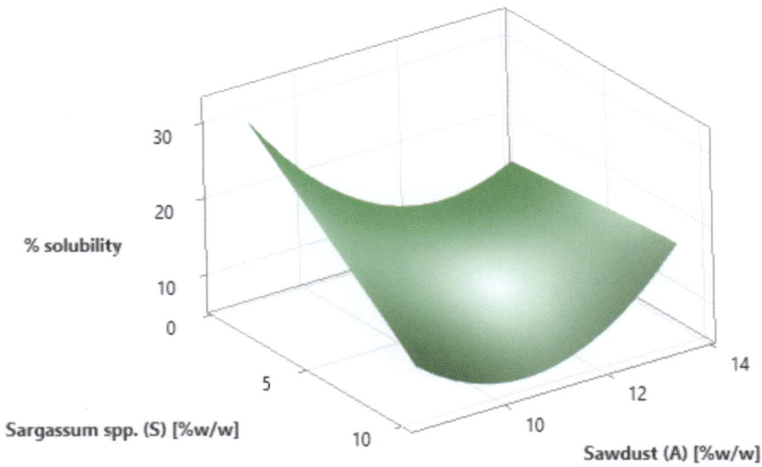

Fig. 2 Response surface for % solubility according to the proposed DOE

(MCC) are distinguished in the spectra, confirming the presence and contribution of each component in the biopolymer matrix. Likewise, the polymer formation typical of this type of material is observed, evidenced by the combination of amorphous and crystalline phases that support the hybrid nature of the obtained bioplastics.

Figure 6 shows the thermograms obtained by thermogravimetric analysis (TGA) of the selected formulations. In this study, C corresponds to the carbon control, A to sawdust, and S to sargassum, while combinations S3A1, S2A3, and S4A3 represent biopolymers with different proportions of sargassum and sawdust. Additionally, sample A2G3 corresponds to a biopolymer formulated with sawdust but without sargassum. The thermal profiles enable the identification of the stages of

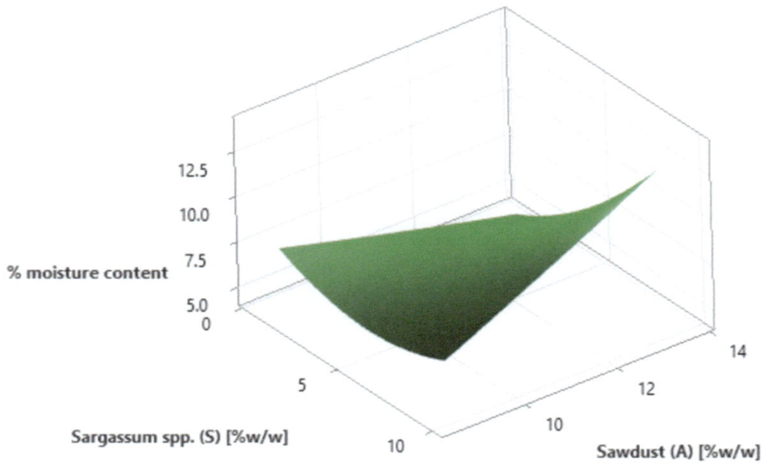

Fig. 3 Response surface for % humidity according to the proposed DOE

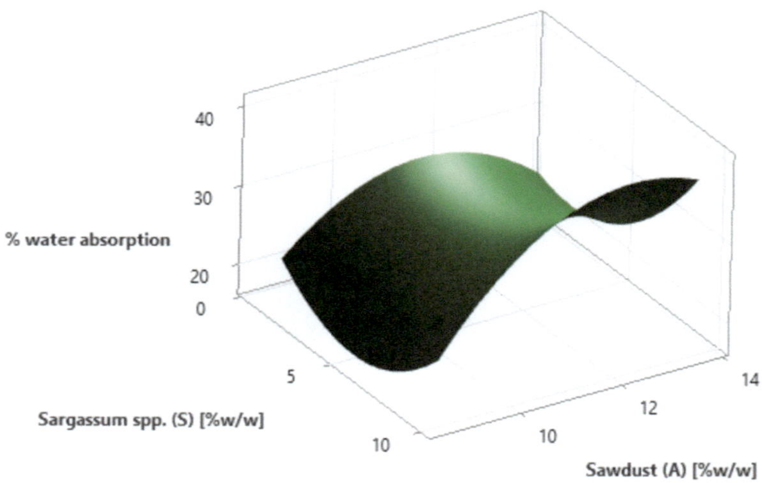

Fig. 4 Response surface for % water absorption according to the proposed DOE

mass loss associated with dehydration, decomposition of lignocellulosic components, and polymer degradation, providing evidence on the thermal stability and degradation behavior of the developed materials.

Figure 7 presents the micrographs obtained by scanning electron microscopy in backscattering mode of samples (a) S4A3, (b) S3A1, and (c) A2G3, selected according to their representative properties. These images illustrate the distribution and phase contrast resulting from the incorporation of sawdust and sargassum into the biopolymer matrix. Also included is a representative EDS spectrum (d) in which

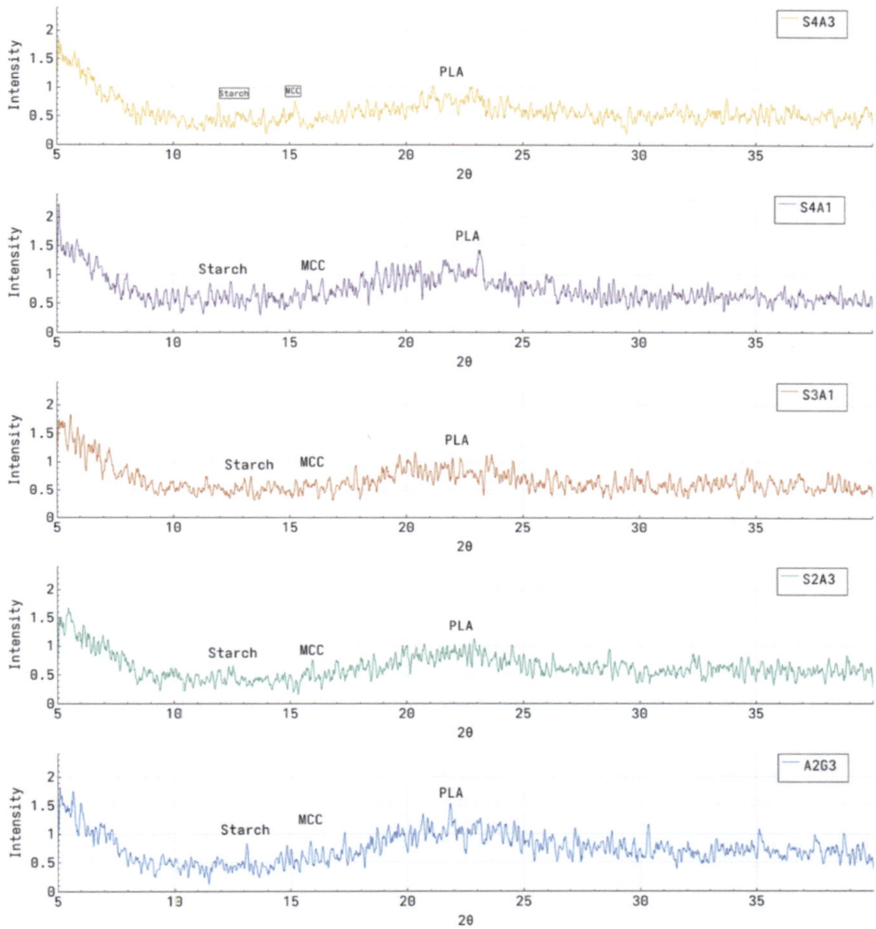

Fig. 5 XRD spectra of samples S4A3, S4A1, S3A1, S2A3, and A2G3 (as a control sample since it does not have sargassum)

the characteristic elements of both biomasses are identified, confirming the presence of mineral components typical of *Sargassum spp.* and elements associated with the lignocellulosic fraction of the sawdust.

Regarding mechanical properties, the biopolymer films exhibited an average ultimate tensile strength (UTS) of 0.032 MPa, with fracture strength values reaching approximately 0.030 MPa. The yield strength was determined to be 0.020 MPa, indicating the relatively brittle nature of the material compared to conventional polymers. Regarding biodegradability, the samples showed an average mass loss of 40% over 20 days under composting conditions, confirming their potential as a sustainable and easily degradable material in natural environments.

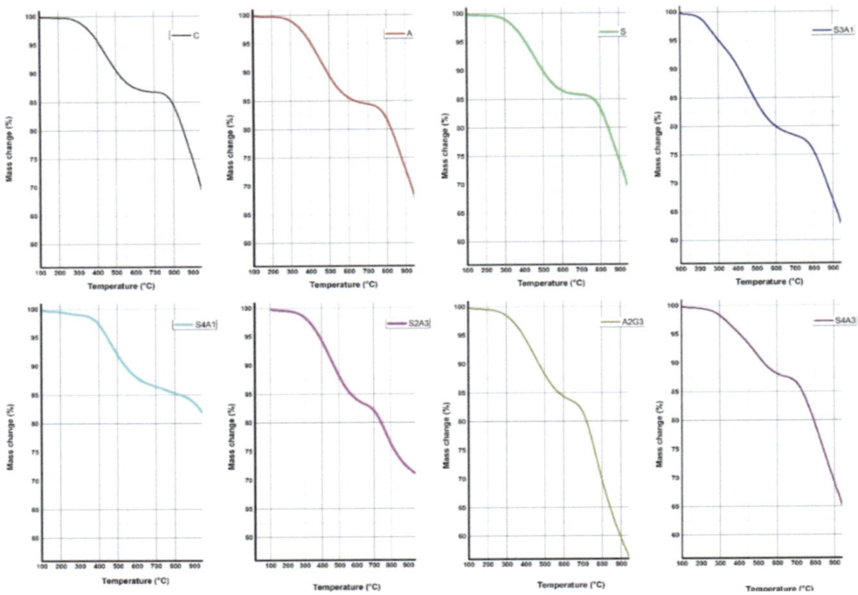

Fig. 6 Thermograms of samples S4A3, S4A1, S3A1, S2A3, and A2G3 (as a control sample since it does not have sargassum)

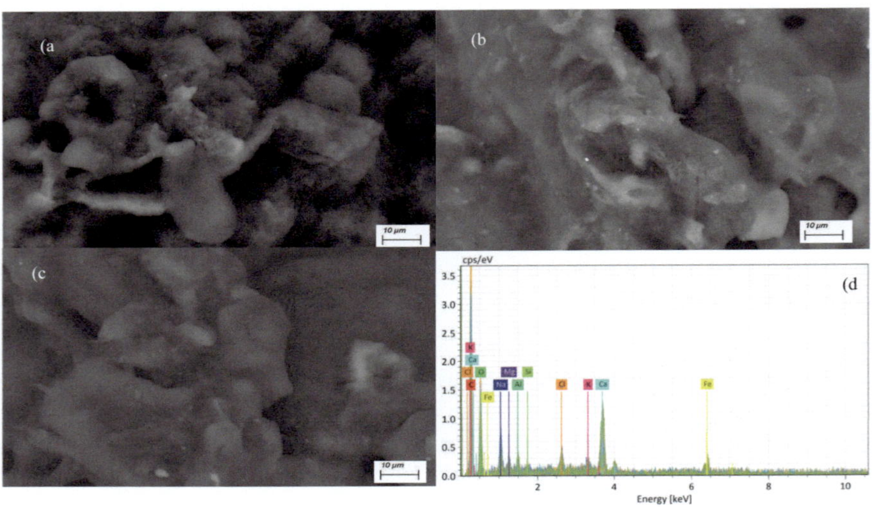

Fig. 7 Micrographs of the samples: **a** S4A3, **b** S3A1, **c** A2G3, and **d** Representative EDS spectrum

Discussion

The results obtained enable us to determine the impact of incorporating Sargassum spp. and sawdust on the physical–chemical, structural, thermal, and mechanical properties of the formulated biopolymers.

First, density (Fig. 1) increased as the proportion of both aggregates increased, with sawdust having the most significant impact on this parameter. This behavior is associated with the lignocellulosic and compact nature of sawdust, which favors denser packing in the polymer matrix.

Regarding water solubility (Fig. 2), it was observed that the percentage increased in formulations with sawdust contents between 10 and 12%, while in lower proportions of both aggregates, the solubility was lower. This result suggests that sawdust acts as a point of heterogeneity within the matrix, facilitating the entry of water and the partial dissolution of soluble components.

Moisture content (Fig. 3) also increased as the proportion of both reinforcements increased, although with a greater dependence on sawdust. In addition, water absorption (Fig. 4) exhibited the same trend, which is attributed to the high number of hydroxyl groups present in both the cellulose in the sawdust and the polysaccharides in the sargassum, promoting interactions with water molecules.

In the structural analysis, XRD diffractograms (Fig. 5) revealed characteristic peaks of starch, PLA, and MCC in all samples, confirming the expected polymer formation. Among them, samples S4A1 and A2G3 showed a higher proportion of signals associated with PLA, suggesting better integration of the components and a greater degree of order in the matrix.

TGA analysis (Fig. 6) showed distinct thermal degradation patterns compared to the controls, reflecting the influence of lignocellulosic fillers on material stability. In particular, sample S3A1 exhibited a broader degradation range, indicating greater thermal resistance and a more complex decomposition profile.

Backscattering SEM micrographs (Fig. 7) confirmed the morphological differences between the formulations. Sample S3A1 exhibited a more homogeneous and polymerized structure, while S4A3 showed the presence of incompletely polymerized particles. The EDS spectrum revealed the presence of elements such as Ca, Fe, K, Si, Cl, and Na, characteristic of Sargassum spp. and sawdust, in addition to a high carbon content attributable to the polymer chains of the matrix.

Regarding mechanical properties, the average maximum tensile strength was 0.032 MPa, with a fracture strength of 0.030 MPa and a yield strength of 0.020 MPa. While these values are lower than those reported for conventional plastics, they reflect the experimental nature of the material and constitute a solid basis for future formulation optimization.

Finally, biodegradability was confirmed by an average mass loss of 40% in 20 days, indicating the potential of these biopolymers as environmentally friendly materials. Although their mechanical properties currently limit their application to structural uses, the results suggest their viability as support materials or in applications where high strength is not required, while also leveraging the potential for recycling waste such as sawdust and sargassum.

Conclusions

This study reports the development of starch-based biopolymer films reinforced with Sargassum spp. (1.3–10% w/w) and sawdust (8.5–13.6% w/w) following a two-factor, four-level factorial design. The optimal formulation, with a 35:16 sawdust: sargassum ratio, exhibited a density of 1.3 g/mL, a 3.5% moisture content, a 19.4% solubility, a 27.3% water absorption, and a 40% biodegradability in 20 days.

Structural analysis (XRD) confirmed the formation of polymers with semi-crystalline characteristics attributed to cellulose and alginates, while TGA demonstrated thermal stability. SEM–EDS micrographs revealed a heterogeneous but cohesive morphology, with elements characteristic of sargassum and sawdust.

In terms of mechanical properties, the films exhibited low tensile strength but greater flexibility, which limits their use in applications requiring high strength. However, they are promising as biodegradable materials for packaging and agricultural applications.

Overall, the results confirm the viability of utilizing waste such as sawdust and sargassum in the production of sustainable biopolymers, contributing to biomass valorization and strengthening the circular economy.

References

1. Sathish Kumar P, Ramalingam S, Arumugam A et al (2024) Biodegradable plastics from marine biomass: a solution to marine plastic pollution. Mar Pollut Bull 207:114234. https://doi.org/10.1016/j.marpolbul.2024.114234
2. Rasyid R, Rahman A, Mustafa M et al (2021) A sustainable approach to green algal bioplastics production from brown seaweeds of Sabah, Malaysia. Biocatal Agric Biotechnol 37:102183. https://doi.org/10.1016/j.bcab.2021.102183
3. Allouch A, Perrin D, Brosse N (2022) Enhancing lignin dissolution and extraction: the effect of surfactants. Ind Crops Prod 189:115844. https://doi.org/10.1016/j.indcrop.2022.115844
4. de Paula R, Silva M, da Silva V et al (2022) Extraction and characterization of two residual lignins from eucalyptus wood. Polymers 32(1):e2022003. https://doi.org/10.1590/0104-1428.20220003
5. Hernández-Carmona G, Freile-Pelegrín Y, Robledo D (2023) Polysaccharides from the Sargassum and brown algae genus: extraction, purification, and their potential therapeutic applications. Mar Drugs 21(1):25. https://doi.org/10.3390/md21010025
6. Bustos C, Acuña S, Rojas C et al (2023) Process design and scale-up study for the production of polyol-based biopolymers from sawdust. J Clean Prod 385:135705. https://doi.org/10.1016/j.jclepro.2022.135705
7. Smith A, Thomas L, Ferreira R et al (2023) Qualitative and quantitative assessment of Sargassum valorization solutions for the Caribbean. J Environ Manag 330:117114. https://doi.org/10.1016/j.jenvman.2023.117114

8. Roldán-San Antonio J, Martín M (2023) Optimal integrated plant for biodegradable polymer production. Comput Chem Eng 170:108131. https://doi.org/10.1016/j.compchemeng.2023.108131
9. Ramírez-López C, Méndez J, González A et al (2023) Sargassum-inspired, optimized calcium alginate bioplastic composites for food packaging. Food Packag Shelf Life 38:101136. https://doi.org/10.1016/j.fpsl.2023.101136
10. El Ouahabi S, Boujaber N, Moumen A et al (2023) Seasonal variation in biochemical and metal content of the invasive seaweed Sargassum muticum along Moroccan Atlantic coast. RegNal Stud Mar Sci 61:102907. https://doi.org/10.1016/j.rsma.2023.102907
11. Ali S, Johnson R, Matthews P et al (2023) The environmental impact and economic feasibility assessment of composite calcium alginate bioplastics derived from Sargassum. J Clean Prod 406:136996. https://doi.org/10.1016/j.jclepro.2023.136996

Influence of Beetle Larvae Oil on the Properties of Cassava Starch-Based Films

M. V. F. Monteiro, D. C. Mendes, S. N. Monteiro, A. C. R. Silva, and V. S. Candido

Abstract This study aimed to develop and characterize polymeric films based on poly(vinyl alcohol) (PVA), cassava starch, and oil extracted from tucumã larvae (*Speciomerus ruficornis*) using the casting technique. Formulations were prepared at different concentrations to evaluate structural, physicochemical, and mechanical properties. Optical microscopy showed that neat PVA exhibited a homogeneous and continuous surface, reflecting good polymer chain cohesion. The addition of starch led to dispersed domains, indicating phase interactions and contributing to structural reinforcement. Incorporation of oil resulted in hydrophobic microdroplets, revealing typical phase separation of emulsified systems. Moisture tests demonstrated that starch-containing films retained more water than neat PVA due to hydroxyl groups available for hydrogen bonding. Oil addition mitigated this effect, reducing moisture at intermediate concentrations. Regarding solubility, all films fully dissolved in water, confirming the hydrophilic nature of the matrices. Tensile tests revealed that starch significantly increased mechanical strength, while oil acted as a plasticizer at higher levels, decreasing rigidity. Overall, the PVA–starch formulation proved most effective for structural reinforcement, whereas oil can be used to modulate flexibility and functional properties. These findings highlight the potential application of these films as biodegradable and active packaging, aligning with sustainability goals and the reduction of conventional plastics.

M. V. F. Monteiro (✉) · D. C. Mendes · A. C. R. Silva · V. S. Candido
Federal University of Pará, Ananindeua Campus/PA, Ananindeua, Brazil
e-mail: manoelviniciuz@gmail.com

D. C. Mendes
e-mail: daiane.mendes@ananindeua.ufpa.br

A. C. R. Silva
e-mail: alissonrios@ufpa.br

V. S. Candido
e-mail: scarpini@ufpa.br

S. N. Monteiro · V. S. Candido
Military Institute of Engineering/RJ, Rio de Janeiro, Brazil
e-mail: snevesmonteiro@gmail.com

H. A. C. Lopera et al. (eds.), *Natural Fibers and Biocomposites 2026*, The Minerals, Metals & Materials Series, https://doi.org/10.1007/978-3-032-13577-3_23

Keyword Biopolymers · Biodegradable packaging · Poly (vinyl alcohol)

Introduction

Reinforced bioplastics are emerging as promising alternatives to conventional plastics, especially in food packaging. Recent studies have shown that the incorporation of reinforcements into biodegradable matrices enhances mechanical and barrier properties, as well as extends the shelf life of food products [1, 2].

The oil from the tucumã seed beetle larva (*Speciomerus ruficornis Germar*) is extracted from the insect in its larval stage. The female beetle lays its eggs on the tucumã seeds, and once hatched, the larvae penetrate and lodge inside the seed, feeding on the kernel and sheltering there until completing their larval development. At the end of this process, the larvae emerge as beetles, leaving behind this temporary casing. However, before the final stage of development, the larvae are collected by local populations for the extraction of oil, which is popularly known as 'larva oil' or 'insect oil' [3].

Cassava starch-based bioplastic films have gained prominence due to their ability to form biodegradable and renewable matrices, with potential applications in active packaging. The incorporation of additives such as essential oils (eugenol), natural extracts, or reinforcements (nanoclays, fibers) has been shown to improve mechanical properties, provide antimicrobial activity, and reduce water vapor permeability, thereby extending the shelf life of food products [4–6].

In addition, polymer-based materials are widely used since they can be derived from natural sources while also meeting specific requirements, such as moisture retention [7]. Poly(vinyl alcohol) (PVA) offers several advantages, including good film-forming properties, thermal stability, biocompatibility, flexibility, and transparency [8]. The presence of hydroxyl groups in both PVA and starch leads to the formation of intermolecular and intramolecular hydrogen bonds, which contributes to the higher water absorption of the films [8]. The water resistance and compatibility of PVA/cassava starch (PVA/CS) films can be further enhanced through the incorporation of components such as oil-based emulsions.

Oil-in-water emulsions are colloidal systems formed by dispersing oil droplets in a continuous phase, generally aqueous, stabilized by emulsifying agents [9, 10]. These emulsions have broad applications in various fields, such as cosmetics, food, and pharmaceuticals, due to their ability to combine hydrophilic and hydrophobic properties within a single product [11].

The present study aimed to develop and characterize polymeric films obtained from the combination of cassava starch and insect oil, using casting and spontaneous emulsification methods. The study investigated the interaction between the polymeric matrix and the oil emulsion at different concentrations, to evaluate the influence of these components on the structural, functional, and physicochemical properties of the films produced.

Materials and Methods

Materials

The materials and reagents used were PVA powder (molecular weight 72,000, viscosity 44.4, with a maximum of 0.5% of other components); insect oil, manually extracted, liquid, slightly yellowish, and non-aromatic; Tween 80 (polysorbate), Chemical Abstracts Service (CAS): 9005–65-6; and commercial cassava starch (Delícia brand).

Methods

Preparation of the Beetle Larval Oil Emulsion

The emulsions were prepared using 40 mL of distilled water, 2 mL of Tween 80 (polysorbate), and 2 mL of insect oil. Initially, the surfactant was solubilized in the oil phase (insect oil), and the mixture was then added dropwise to the aqueous phase under continuous magnetic stirring for 20 min at room temperature. The formulations were prepared at the Materials Characterization Laboratory of the Federal University of Pará, Ananindeua Campus.

Preparation of Control PVA Film

The synthesis of poly(vinyl alcohol) (PVA) films and cassava starch (CS) films was carried out based on the protocol described by [12], with the necessary methodological adaptations. The films were individually obtained using the casting technique. For the PVA + H_2O system formulation, 1 g of PVA was dissolved in 100 mL of distilled water,the solution was subjected to magnetic stirring for 5 min at 45 °C and subsequently poured into Petri dishes. The drying step was performed in a forced-air circulation oven (model SX 1.2, Sterilifer) for 24 h at 45 °C. All film preparations were conducted at the Materials Characterization Laboratory of the Federal University of Pará, Ananindeua Campus.

Preparation of Cassava Starch/PVA Blends

Cassava starch/poly(vinyl alcohol) (PVA) blends were obtained using the casting technique. Initially, 0.7 g of PVA was solubilized in 70 mL of distilled water at 45 °C under magnetic stirring for 5 min. Subsequently, 0.3 g of cassava starch and 30 mL of distilled water were added, and the mixture was kept under stirring at 45 °C for an additional 15 min. The resulting solution was poured into Petri dishes and dried in a non-circulating air oven (Sterifiller) for 24 h at 45 °C. Film preparation was carried out at the Materials Characterization Laboratory of the Federal University of Pará, Ananindeua Campus.

The blends containing insect oil emulsion were prepared following the same procedure described in item 2.2.3, with the addition of 0.1%, 0.5%, 1%, and 5% (v/v) of the oil emulsion. The formulations were maintained under magnetic stirring at 45 °C for 20 min. Subsequently, the solutions were poured into Petri dishes and dried in a forced-air oven (model SX 1.2, Sterilifer) at 45 °C for 24 h. The experimental design adopted was based on the study by [13].

Results and Discussion

Optical Microscopy of Polymeric Films

The optical micrographs obtained for the polymeric films allow evaluation of matrix uniformity and interactions between the added components. According to Fig. 1, the neat PVA film tends to exhibit a homogeneous and continuous surface, reflecting the well-organized polymer chains. With the introduction of starch, regions of higher contrast are commonly observed, associated with the presence of starch domains, whose proper dispersion can contribute to mechanical reinforcement. Recent studies show that uniform distribution of starch in PVA matrices promotes hydrogen bonding and improves structural properties [14].

On the other hand, films containing oil exhibit morphologies with dispersed globular domains, typical of polymeric systems with hydrophobic additives. This heterogeneity can act as a plasticizer, promoting greater flexibility, but in excess it compromises interfacial cohesion and reduces tensile strength. Similar observations have been reported in PVA–essential oil films, where optical microscopy reveals phase separation correlated with decreased mechanical rigidity [15]. Thus, microscopy confirms the direct relationship between the formed microstructure and the functional performance of the polymeric films.

Moisture Content and Solubility Results of the Polymeric Films

Table 1 shows the moisture content results of the polymeric films fabricated with PVA, starch, and larva oil.

The incorporation of cassava starch into PVA led to an increase in the moisture content of the films (14.04%) compared to neat PVA (10.79%), a behavior attributed to the high hydrophilicity of starch, which contains hydroxyl groups capable of forming hydrogen bonds with water, thereby enhancing the material's water retention capacity. This effect has been reported in other studies, which indicate higher

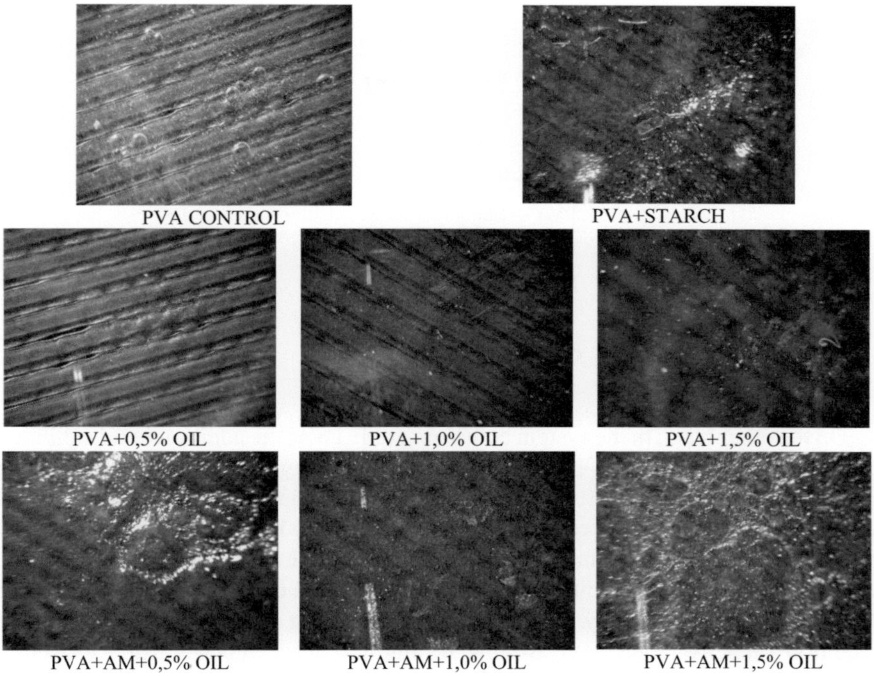

Fig. 1 Optical microscopy of the fabricated films

Table 1 Moisture content results of the fabricated films

Composition	PU (%)
PVA control	10,79
PVA + 0,5% de oil	11,91
PVA + 1,0% de oil	10,11
PVA + 1,5% de oil	11,42
PVA + Starch	14,04
PVA + Starch + 0.5% oil	13,05
PVA + Starch + 1.0% oil	12,64
PVA + Starch + 1.5% oil	13,05

moisture absorption in PVA/starch films due to the greater availability of hydrophilic sites for interaction with water molecules [16, 17]. The presence of larva oil mitigated this effect, reducing moisture content in all starch-containing formulations, with a notable decrease observed in PVA + Starch + 1.0% oil (12.64%). This reduction can be explained by the hydrophobic nature of the lipids in the oil, which form partial barriers to water diffusion within the matrix, decreasing the number of active sites available for water adsorption.

Table 2 Solubility results of the fabricated films

Composition	S (%)
PVA control	100
PVA + 0,5% de oil	100
PVA + 1,0% de oil	100
PVA + 1,5% de oil	100
PVA + Starch	100
PVA + Starch + 0.5% oil	100
PVA + Starch + 1.0% oil	100
PVA + Starch + 1.5% oil	100

Source Authors, 2025

On the other hand, in the system containing only PVA, the addition of oil exhibited a nonlinear behavior, with moisture content decreasing at 1.0% oil (10.11%) but increasing at 0.5% and 1.5%. This result may be associated with heterogeneous oil dispersion, where low concentrations are insufficient to form effective barriers, and higher concentrations can induce defects or phase separation regions that facilitate water entrapment [17]. Additionally, the presence of oil can alter the microstructure and polymer chain mobility, interfering with the drying process and the final water retention capacity of the film. Similar results have been reported for PVA/starch composites modified with hydrophobic additives, showing reduced water absorption up to an optimal oil concentration, followed by stabilization or slight increase at higher levels [16]. Table 2 presents the solubility results of the polymeric films.

The results indicate that all films exhibited 100% solubility, regardless of formulation, demonstrating the highly water-soluble nature of the polymeric matrix. This behavior is expected for PVA-based films, since the polymer is highly hydrophilic and contains hydroxyl groups capable of strongly interacting with water, leading to complete dissolution when immersed in an aqueous medium [16]. The addition of cassava starch did not alter this characteristic, as starch is also hydrophilic and water-soluble when gelatinized, contributing to the maintenance of high solubility. The absence of a significant effect from larva oil suggests that the concentrations used (0.5–1.5%) were not sufficient to generate hydrophobic barriers capable of limiting water diffusion into the matrix or preventing film disintegration.

The fact that all films exhibited complete solubility can be advantageous for applications requiring rapid biodegradability or controlled release of active compounds in aqueous media. However, for applications where water resistance or use in humid environments is necessary, chemical modification strategies would be required, such as PVA crosslinking or the use of hydrophobic agents in higher proportions or with improved dispersion to reduce the dissolution rate [17]. Recent studies report that treatments with boric acid, citrate, or aldehydes promote PVA chain crosslinking, decreasing solubility and improving water stability, which could be an alternative for future developments with this type of biocomposite [18].

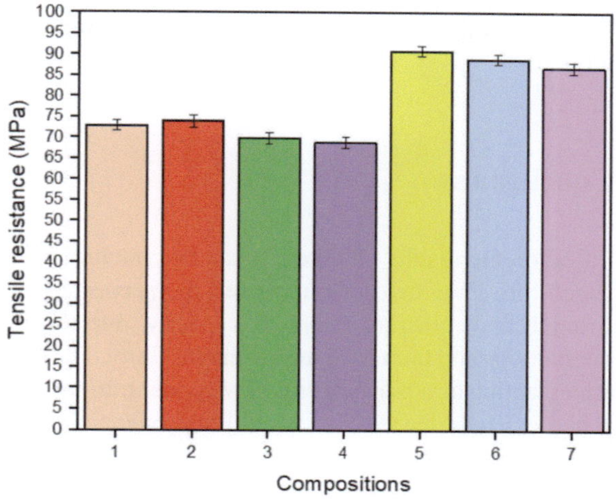

Fig. 2 Tensile strength of the fabricated films

Tensile Strength Results of the Polymeric Films

Figure 2 shows the tensile strength graph of the composites, with the compositions represented by numbers, where: 1 corresponds to the neat PVA film, 2 to PVA + 0.5% oil, 3 to PVA + 1.0% oil, 4 to PVA + 1.5% oil, 5 to PVA + starch, 6 to PVA + starch + 0.5% oil, 7 to PVA + starch + 1.0% oil, and finally, 8 to PVA + starch + 1.5% oil.

The tensile strength results highlight the contrasting roles of oil and starch in modifying PVA properties. Neat PVA exhibited intermediate strength, while the addition of small amounts of oil (0.5%) led to a slight increase, possibly due to its partial plasticizing effect, which enhances segmental mobility without compromising matrix cohesion. However, increasing the oil concentration (1.0 and 1.5%) significantly reduced strength, a result consistent with the literature showing that higher additions of essential or vegetable oils can induce morphological heterogeneity and impair load transfer between polymer chains [15, 19].

On the other hand, the incorporation of cassava starch provided substantial reinforcement, reaching strength values higher than the neat PVA, which can be explained by the strong interaction between the hydroxyl groups of starch and PVA through hydrogen bonding, in addition to the increased crystallinity and rigidity of the matrix. This behavior has been widely reported in biodegradable PVA–starch films, which show significant improvements in mechanical performance [14, 20]. In the formulations combining starch and oil, high strength values were maintained, although slightly lower than those of starch alone, highlighting the simultaneous effects of structural reinforcement and plasticization. Therefore, the analysis suggests that the PVA–starch formulation represents the most promising condition for applications requiring superior mechanical performance, whereas the combination with oil may be strategically used when aiming to balance strength with flexibility and functional properties.

Conclusions

Based on the results obtained, it can be concluded that the incorporation of cassava starch into PVA matrices improved the properties of the films, particularly tensile strength and moisture retention capacity, due to the formation of hydrogen-bond interactions and increased matrix crystallinity. The addition of larva oil showed a dual effect, acting as a plasticizer at low concentrations but reducing mechanical strength at higher levels due to the structural heterogeneity observed under optical microscopy. The combined analysis demonstrates that the PVA–starch blend represents the most promising formulation for structural reinforcement, while the associated use of oil can be explored to modulate flexibility and functional properties. Thus, the developed films show potential application in biodegradable and active packaging, combining mechanical performance, hydrophilic characteristics, and the possibility of functional modification through natural additives.

References

1. Putri AG, Khandeparkar PN, Cruz G et al (2025) Sustainable biodegradable tapioca starch films enhanced with basil oil, carboxymethyl cellulose, and citric acid for functional food packaging applications. Front Sustain Food Syst. https://doi.org/10.3389/fsufs.2025.1610769.
2. Siddiqui SA, Yang X, Deshmukh RK, Gaikwad KK, Bahmid NA, Castro-Muñoz R (2024) Recent advances in reinforced bioplastics for food packaging–a critical review. Int J Biol Macromol 263(Part 2):130399. https://doi.org/10.1016/j.ijbiomac.2024.130399
3. Rocha TT, Tavares-Martins ACC, Lucas FCA, Martins RCC (2014) Potencial terapêutico e composição química do óleo do bicho do tucumã (Astrocaryum vulgare Mart) utilizado na medicina popular. Scientia Plena 10(11)
4. Ishara GKK, Koliyabandara PA, Samarakoon G (2024) Eco-friendly bio-composite sheets: a study on the utilization of banana peels, cassava starch, and banana stem fibers. Front Sustain. https://doi.org/10.3389/frsus.2024.1410986
5. Santos JF, Silva CWC, Silva BPG et al (2025) Enhancing cassava starch bioplastics with Vismia guianensis alcoholic extract: characterization with potential applications. Polymers 17(3):0419. https://doi.org/10.3390/polym17030419
6. Silveira YDO, Franca AS, Oliveira LS (2025) Cassava waste starch as a source of bioplastics: development of a polymeric film with antimicrobial properties. Foods 14(1):0113. https://doi.org/10.3390/foods14010113
7. Paranhos SB et al (2022) Chitosan membrane containing copaiba oil (Copaifera spp.) for skin wound treatment. Polymers 14(1):35
8. Ghadiri N et al (2023) Physicochemical properties and antioxidant activity of polyvinyl alcohol orally disintegrating films containing sweet almond oil nanoemulsion. J Food Meas Charact 1–15
9. Maurya A et al (2021) Essential oil nanoemulsion as eco-friendly and safe preservative: bioefficacy against microbial food deterioration and toxin secretion, mode of action, and future opportunities. Front Microbiol 12:751062

10. Pandey VK et al (2024) Bioactive properties of clove (Syzygium aromaticum) essential oil nanoemulsion: a comprehensive review. Heliyon
11. Wei HN et al (2024) Characteristics of corn starch/polyvinyl alcohol composite film with improved flexibility and UV shielding ability by novel approach combining chemical cross-linking and physical blending. Food Chem 140051
12. Shrestha B et al (2023) Starch/polyvinyl alcohol (PVA) blend bioplastics: synthesis and physicochemical properties. J Nepal Chem Soc 43(2):103–109
13. Altaf F et al (2021) Synthesis and characterization of PVA/starch hydrogel membranes incorporating essential oils aimed to be used in wound dressing applications. J Polym. Environ. 29:156–174
14. Liu H et al (2024) Development and characterization of starch/PVA blend films for packaging applications. Carbohydr Polym. 320:121308
15. Yildirim S et al (2023) Biopolymer films with essential oils: mechanical and functional performance. Food Packag Shelf Life 37:101093
16. Abedi-Firoozjah R et al (2022) PVA/starch films: an updated review of their preparation, characterization, and diverse applications in the food industry. Polym. Test. 111:107679
17. Wei X et al (2023) Intermolecular interactions between starch and polyvinyl alcohol for improving mechanical properties of starch-based straws. Int J Biol Macromol 225:1492–1502
18. Khan M et al (2023) Crosslinked polyvinyl alcohol-based films: water resistance and biodegradability improvements. J Appl Polym Sci 140(12):e53678
19. Sothornvit R, Krochta JM (2000) Plasticizer effect on oxygen permeability of edible films. J Agric Food Chem 48(12):6298–6302
20. Mali S, Grossmann MVE, Yamashita F (2005) Starch/poly(vinyl alcohol) films: mechanical properties and microstructure. Carbohydr Polym 60:283–289
21. Lima TP, Passos MF (2021) Skin wounds, the healing process, and hydrogel-based wound dressings: a short review. J Biomater Sci Polym 32(14):1910–1925

Natural Fiber from the Amazon: Thermal and Morphological Characterization of Mucajá (Acrocomia Aculeata)

Damares da Cruz Barbosa Nascimento, Sérgio Neves Monteiro, and Verônica Scarpini Candido

Abstract The development of innovative materials has driven the search for alternative and sustainable raw materials that enhance performance while maintaining environmental responsibility. In this context, natural fibers with low density, favorable mechanical properties, and biodegradability have emerged as promising candidates. This study aimed to investigate the thermal and morphological characteristics of Mucajá fiber. The fibers were characterized using Thermogravimetric Analysis (TGA), Differential Scanning Calorimetry (DSC), and Scanning Electron Microscopy (SEM). TGA results demonstrated thermal stability up to approximately 262 °C, with complete degradation occurring near 700 °C. DSC analysis revealed an endothermic peak at 77.15 °C ($\Delta H = 147$ J) and an exothermic peak with enthalpy of 54 J. SEM micrographs showed surface irregularities and the presence of pores along the longitudinal section of the fiber. These findings suggest that Mucajá fibers exhibit promising characteristics for application in various sectors, particularly in the development of composite materials.

Keywords Natural fibers · Morphological characterization · Composite materials

Introduction

The search of biodegradable, sustainable, and cost-effective materials has significantly increased scientific and industrial interest in natural alternatives to conventional synthetic materials [1]. Current research emphasizes the use of raw resources that satisfy environmental requirements while simultaneously providing

D. da Cruz Barbosa Nascimento (✉) · V. S. Candido
Materials Science and Engineering Program, Federal University of Pará–UFPA, Ananindeua, Brazil
e-mail: damares.barbosa62@gmail.com

D. da Cruz Barbosa Nascimento · S. N. Monteiro
Materials Science Program, Military Institute of Engineering–IME, Rio de Janeiro, Brazil

H. A. C. Lopera et al. (eds.), *Natural Fibers and Biocomposites 2026*, The Minerals, Metals & Materials Series, https://doi.org/10.1007/978-3-032-13577-3_24

the mechanical performance essential for commercial viability [2]. In this context, composite materials—engineered through the combination of natural and synthetic constituents—are designed to establish a balance between environmental and economic demands while also addressing broader social needs. By integrating two or more components, these systems exhibit synergistic effects, resulting in unique characteristics and properties not attainable by the individual phases alone [3, 4].

Among renewable reinforcements, natural fibers (NFLs) stand out as highly promising due to their abundance, low density, favorable mechanical performance, excellent wear resistance, and inherent biodegradability [5–7]. Their hierarchical architecture is complex, comprising layers, sublayers, and microfibrils, and is mainly composed of cellulose, lignin, hemicellulose, and pectin. The relative proportions and structural arrangements of these constituents govern the mechanical, thermal, and physicochemical properties of the fibers [8, 9].

Despite these advantageous attributes, the application of NFLs under elevated temperatures remains a critical challenge, as these materials undergo a range of physical and chemical transformations when exposed to thermal environments. Fibers such as water hyacinth, reed, roselle, and sisal, although all natural in origin, display marked differences in thermal stability, decomposition temperatures, and microstructural features, which ultimately define their distinct performance profiles [10, 11].

Pinheiro et al. [11, 12] investigated the mechanical, chemical, and thermal behavior of a natural fiber from the Amazon and reported that, in addition to exhibiting good mechanical strength, periquiteira fibers display thermal stability up to approximately 250 °C. Similarly, Demóstenes et al. [12, 13] studied buriti fibers and observed thermal stability around 200 °C.

In Brazil, several plant species and their fibers remain underexplored by the technical-scientific and industrial community with regard to applications in engineering materials. A notable example is the natural fiber of mucajá (*Acrocomia aculeata*), from the Arecaceae family, which, despite being native to several Brazilian regions such as Minas Gerais and São Paulo, has not yet been systematically investigated in this context. Traditionally employed in the manufacture of baskets and fishing lines, this fiber presents characteristics that suggest untapped potential for industrial applications, particularly in the development of composite materials [14–17]. Based on this context, the present work aims to investigate the thermal characteristics of mucajá natural fiber, seeking to assess its potential for future use in composite materials [5].

Materials and Methods

The Mucajá fiber was extracted from the leaves of the *Acrocomia aculeata* plant, located in the Lower Tocantins region, specifically in the municipality of Moju, Jutaiteua community, Cairari district, Pará, Brazil. The extraction took place during the Amazonian summer. Fiber separation was performed manually using a metallic tweezer.

Thermogravimetry (TGA) and Differential Scanning Calorimetry (DSC)

The thermal characterization of Cotia fibers was performed using thermogravimetric analysis (TG) and differential scanning calorimetry (DSC) at the Geosciences Laboratory of the Federal University of Pará (UFPA). The fibers were cut and macerated in a porcelain mortar and pestle in order to reduce the particle size of the sample. The characterization was performed using NEXTA STA300 equipment, with simultaneous thermal analysis (STA), a heating rate of 10 °C/min, starting from room temperature (30 °C) up to 600 °C.

Scanning Electron Microscopy (SEM)

The images obtained by Scanning Electron Microscopy were obtained at the Institutional Scanning Electron Microscopy Laboratory of the Emílio Goeldi Museum of Pará, using a TESCAN Mira3 electron microscope with a FEG (field emission gun) electron gun, OXFORD INSTRUMENTS, with 55,000 × magnification. The samples were metallized with gold (Au) for 2″30″, depositing a thin film on the sample surface. The images were generated by secondary electron detection, using a voltage of 5.0 kV.

Results and Discussion

Thermogravimetry (TGA)

Figure 1 shows the thermogravimetric (TG) curve and the derivative (DTG) obtained for Mucajá fiber.

Thermogravimetric analysis (TGA) of mucajá fiber revealed a multistage degradation profile with three well-defined mass-loss events. The first stage, observed at approximately 94 °C, accounted for an 8.69% reduction in mass and is attributed to the evaporation of physically adsorbed and chemically bound water, reflecting the hygroscopic nature of the fiber [20–22].

The second stage, occurring between 262 and 320 °C with a mass loss of 47.37%, corresponds predominantly to the thermal degradation of hemicellulose and the

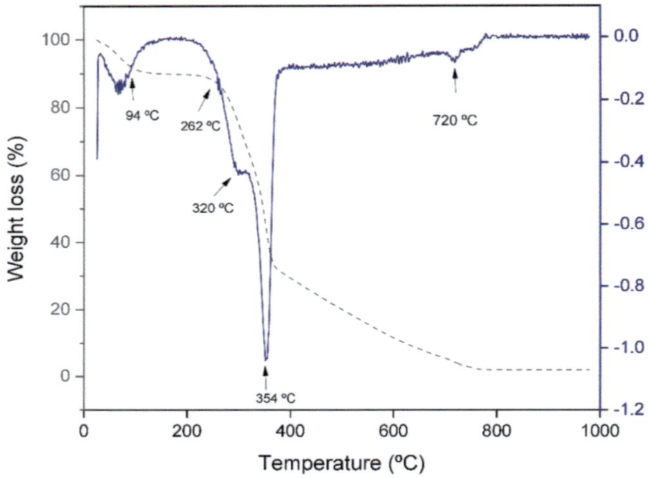

Fig. 1 Thermogravimetry of Mucajá fiber

cleavage of glycosidic linkages, processes typically associated with the onset of structural destabilization in lignocellulosic fibers [17].

The third stage, centered at around 354 °C and responsible for a 43.94% mass reduction, is mainly related to the decomposition of cellulose, accompanied by the breakdown of other polymeric constituents within the fiber matrix [22, 23].

At higher temperatures, the residual material exhibited gradual degradation, culminating in complete decomposition near 720 °C. This final stage indicates the collapse of the lignocellulosic framework, providing evidence of the overall thermal stability and decomposition pathway of mucajá fiber.

The distinct thermal events observed in mucajá fiber emphasize its unique degradation behavior compared to other natural fibers. In particular, the pronounced mass loss around 354 °C, associated with cellulose decomposition, underscores the importance of carefully controlling processing temperatures in composite applications. Such knowledge is crucial for optimizing thermal treatments and ensuring the structural integrity and performance of materials reinforced with this fiber.

Differential Scanning Calorimetry (DSC)

Figure 2 shows the differential scanning calorimetry (DSC) curve of the Mucajá fiber.

Differential scanning calorimetry (DSC) of mucajá fibers revealed an endothermic peak at approximately 77.15 °C, associated with an enthalpy change (ΔH) of 147 J, which corresponds to the release of structurally bound water within the fiber matrix. An additional exothermic event of 54 J/g was also identified, possibly related to

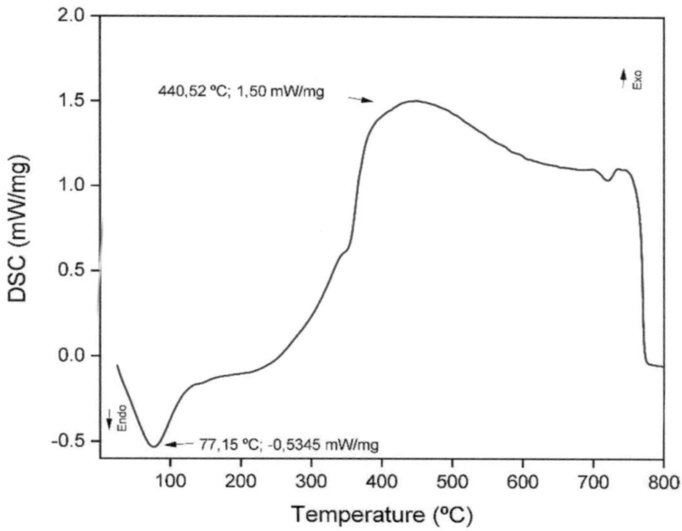

Fig. 2 Differential scanning calorimetry (DSC) curve of Mucajá fiber

secondary structural rearrangements or the onset of oxidative reactions. The correlation between the fiber's hygroscopic nature and the endothermic event suggests that the early stages of thermal degradation initiate around 100 °C, marking the progressive destabilization of the fiber.

When compared with other natural fibers commonly used in composites, such as curauá, sisal, and buriti, mucajá demonstrates distinctive thermal behavior. While curauá and sisal fibers typically exhibit endothermic water-loss peaks below 100 °C and maintain stability up to approximately 200–250 °C, mucajá shows an earlier onset of structural destabilization. Buriti fibers, for instance, are stable up to around 200 °C, whereas mucajá exhibits exothermic signals even before this range, indicating a different structural composition and degradation pathway. These differences reinforce the uniqueness of mucajá fiber and highlight the importance of investigating its thermal response, since its atypical behavior may influence both processing conditions and performance in composite applications.

A thermal event around 350 °C, likely related to cellulose degradation, highlights the unique thermal behavior of mucajá fiber. This transition can directly impact processing conditions and the performance of composites reinforced with this fiber.

Scanning Electron Microscopy (SEM)

The morphological analysis of the Mucajá fiber is presented in Fig. 3.

Fig. 3 SEM analysis of the fiber cross-section

Figure 3 shows that Mucajá fibers present rough and irregular surfaces along their longitudinal axis, with evident pores and aligned microfibrillar bundles distributed throughout the fiber length. Such morphological features, while potentially detrimental to intrinsic mechanical strength due to stress concentration points, are advantageous for composite applications, as they enhance adhesion and mechanical interlocking with polymer matrices [11, 23, 24]. Manimaran et al. [17] similarly reported porous and irregular surfaces in natural fibers, emphasizing their dual role in mechanical performance and interfacial bonding. In agreement, Ding et al. [25] highlighted that surface constituents such as waxes, pores, and irregularities can contribute positively to fiber–matrix adhesion by promoting interfacial anchoring.

Beyond surface roughness, the microstructural arrangement of the mucajá fibers, including the orientation and density of microfibrillar bundles, plays a critical role in determining both mechanical behavior and interfacial performance in composites. Aligned microfibrils can enhance tensile load transfer along the fiber axis, while the presence of pores and surface irregularities increases the effective surface area available for matrix adhesion. This combination not only improves mechanical interlocking but may also influence the bonding of polymer chains to the fiber, leading to stronger interfacial adhesion. Furthermore, the heterogeneous morphology could affect stress distribution under load, promoting energy dissipation at the fiber matrix interface, a mechanism often observed in lignocellulosic fiber-reinforced composites [17, 25].

Mucajá fibers exhibit distinctive thermal and morphological characteristics that make them promising for composite applications. Their multistage thermal degradation, along with rough and porous surfaces containing aligned microfibrillar bundles, influences both processing behavior and interfacial performance. These features can enhance matrix adhesion and mechanical interlocking, supporting effective load transfer in composites and highlighting the potential of mucajá fibers as a versatile reinforcement material.

Conclusion

The characterizations performed on Mucajá fiber revealed the following:

- Fourier Transform Infrared Spectroscopy (FTIR) analysis showed absorption bands characteristic of natural fibers;
- Thermogravimetric analysis (TGA) indicated that Mucajá fiber exhibits greater thermal stability around 320 °C;
- Differential Scanning Calorimetry (DSC) revealed both endothermic and exothermic peaks, with the endothermic peak showing an enthalpy change of approximately 145 J/g;
- Thermal analyses by TGA and DSC demonstrated structural water loss occurring near and above 100 °C;
- Morphological analysis showed that the fibers have rough surfaces with imperfections and porosity;
- The study of this novel natural fiber indicates its potential for the development of new polymer matrix composite materials with possible industrial applications.

Acknowledgements The authors acknowledge the Graduate Program in Science and Engineering of Materials (PPGCEM), the Research Group in Composite Materials (GPMCOMP), the Coordination for the Improvement of Higher Education Personnel (CAPES), and the Federal University of Pará (UFPA).

References

1. Neto JQH, Aguiar R, Lima R, Cavalcanti D, Banea MD (2022) Revisão dos avanços recentes em compósitos poliméricos híbridos reforçados com fibras naturais. J Renew Mater. 10(3):2164–6341

2. Andrew JJ, Dhakal HN (2022) Compósitos sustentáveis de base biológica para aplicações avançadas: tendências recentes e oportunidades futuras – Uma revisão crítica. Compos Parte C Acesso Aberto. 7:100220. https://doi.org/10.1016/j.jcomc.2021.100220
3. Nayak SY, Mohamed THS et al (2022) Potencial de fibras naturais em compostos para aplicações balísticas – uma revisão. J Nat Fibras. 19(5):1648–1658. https://doi.org/10.1080/15440478.2020.1787919
4. Kumar S, Manna A, Rakesh D (2022) Uma revisão sobre aplicações de compósitos reforçados com fibras naturais (NFRCs). Mater Today Proc. 50(5):1632–1636. https://doi.org/10.1016/j.matpr.2021.09.131
5. Asyraf MRM, Ishak MR, Sapuan SM, Yidris N, Ilyas RA, Rafidah M et al (2020) Aplicação potencial de compósitos verdes para componente de braço transversal em torre de transmissão: uma breve revisão (portuguese). Int J Polym Sci. 2020:1–15. https://doi.org/10.1155/2020/8878300
6. Haris NIN, Mohamad Zaki Hassan RA et al (2022) Propriedades mecânicas dinâmicas de compósitos poliméricos híbridos reforçados com fibras naturais: uma revisão. J Mater Res Technol 19:167–82. https://doi.org/10.1016/j.jmrt.2022.04.155
7. Sabaruddin FA, Paridah MT, Sapuan SM, Ilyas RA, Lee SH, Abdan K et al (2020) Efeitos da nanocelulose não branqueada e branqueada na térmica e inflamabilidade de bionanocompósitos de polímero híbrido com núcleo de kenaf reforçado com polipropileno. Polímeros. 13:116. https://doi.org/10.3390/polym13010116
8. Asrofi M, Sapuan SM, Ilyas RA, Ramesh M (2020) Característica de bioplásticos compostos de amido de tapioca e fibra de bagaço de cana-de-açúcar: efeito do tempo de duração da ultrassonicação (tipo banho). Mater Today Proc. 46:1626–1630. https://doi.org/10.1016/j.matpr.2020.07.254
9. Asim M, Paridah MT, Chandrasekar M et al (2020) Estabilidade térmica de fibras naturais e seus compósitos poliméricos. Iran Polym J 29:625–648. https://doi.org/10.1007/s13726-020-00824-6
10. Kavitha SA, Priya RK, Arunachalam KP, Avudaiappan S, Maureira-Carsalade N, Roco-Videla Á (2023) Investigação sobre propriedades de novas fibras celulósicas de raiz de Zea mays tratadas com álcali e cruas para compósitos poliméricos. Polímeros. 15:1802. https://doi.org/10.3390/polym15071802
11. Pinheiro MA, Ribeiro MM, Rosa DLS, Nascimento DDCB, da Silva ACR, dos Reis MAL, Monteiro SN, Candido VS (2023) Periquiteira (*Cochlospermum orinocense*): a promising Amazon fiber for application in composite materials. Polymers 15(9):2120. https://doi.org/10.3390/polym15092120
12. Demosthenes LCDC, Luz FSD, Nascimento LFC, Monteiro SN (2022) Buriti fabric reinforced epoxy composites as a novel ballistic component of a multilayered armor system. Sustainability 14:10591. https://doi.org/10.3390/su141710591
13. Vianna SA, Campos-Rocha A (2020) *Acrocomia.* In: Flora do Brasil 2020. Jardim Botânico do Rio de Janeiro. https://floradobrasil2020.jbrj.gov.br/FB15663
14. Vianna SA, Campos-Rocha A (2024) *Acrocomia.* In: Flora e Funga do Brasil. Jardim Botânico do Rio de Janeiro. Accessed 22 Mar 2024 https://floradobrasil.jbrj.gov.br/FB15663
15. Deeraj BDS, Harikrishnan R, Jayan JS, Saritha A, Joseph K (2020) Enhanced visco-elastic and rheological behavior of epoxy composites reinforced with polyimide nanofiber. Nano-Structures & Nano-Objects 21:10042
16. Dalmis R, Kilic GB, Seki Y et al (2020) Characterization of a novel natural cellulosic fiber extracted from the stem of *Chrysanthemum morifolium.* Cellulose 27:8621–8634. https://doi.org/10.1007/s10570-020-03385-2
17. Manimaran P, Senthamaraikannan P, Sanjay MR, Marichelvam MK, Jawaid M (2018) Study on characterization of *Furcraea foetida* new natural fiber as composite reinforcement for lightweight applications. Carbohydr Polym 181:650–658
18. Kathirselvam M, Kumaravel A, Arthanarieswaran VP, Saravanakuma SS (2019) Characterization of cellulose fibers in *Thespesia populnea* barks: influence of alkali treatment. Carbohyd Polym 217:178–189. https://doi.org/10.1016/j.carbpol.2019.04.063

19. Liu Y, Jia W, Yang Q, Ma Y et al (2022) Physical, mechanical, and environmental properties of corn stalk fiber reinforced braking composites prepared by wet granulation. J Nat Fibers 19(16):14515–14524
20. Moshi AM, Ravindran D, Sundara SR, Padma SR, Indran S, Divya D (2020) Characterization of natural cellulosic fiber extracted from *Grewia damine* flowering plant's stem. Int J Biol Macromol 164:1246–1255. https://doi.org/10.1016/j.ijbiomac.2020.07.225
21. Vijay R, Kumaran P, Pottathil S, Jafrey DJD, Gunda Y et al (2022) Influence of *Parthenium hysterophorus* and *Impomea pes-caprae* fibers stacking sequence on the performance characteristics of epoxy composites. J Nat Fibers 19(12):4456–4466
22. Oqla Faris MAL, Mohammed TH, Mu'ayyad MAL (2023) Thermal stability analysis and mechanical performance of Mediterranean lignocellulosic fiber reinforced polypropylene sustainable composites. J Ind Eng Chem. https://doi.org/10.1016/j.jiec.2023.12.053
23. Khan A, Vijay R, Singaravelu DL, Sanjay MR, Siengchin S, Verpoort F, Asiri AM (2021) Extraction and characterization of natural fiber from *Eleusine indica* grass as reinforcement of sustainable fiber reinforced polymer composites. J Nat Fibers 18:1742–1750
24. Ferreira DCDON, Ferreira ADS, Monteiro SN (2018) Weibull analysis of tensile tested piassava fibers with different diameters. *Matéria* **23**:e-12254
25. Ding L, Xiaoshuai H, Lihua C et al (2022) Characterization of natural fiber from manau rattan (*Calamus manan*) as a potential reinforcement for polymer-based composites. J Bioresour Bioprod 7(3):190–200. https://doi.org/10.1016/j.jobab.2021.11.002

Tensile, Flexural, and Izod Impact Behavior of Polymeric Composites Reinforced with Varying Matamata Fiber Contents for Eco-friendly Applications in Furniture Design

Diemison Lira Santa Rosa, Marcelo Oliveira de Lima Júnior, Felipe Perissé Duarte Lopes, Noan Tonini Simonassi, Sérgio Neves Monteiro, Alisson Clay Rios da Silva, and Verônica Scarpini Candido

Abstract Currently, there is significant interest for new Amazonian fibers as reinforcements compatible with polymer matrices, owing to their low density and biodegradability. This article evaluates the mechanical behavior of polyester composites with 10, 20, and 30 vol% matamata fiber contents, manufactured by hand lay-up molding and tested for tensile, flexural, and Izod impact properties, and examined using Scanning Electron Microscopy (SEM). The greatest improvements in properties were observed at 30 vol%, with a specific strength of 39.08 ± 2.62 MPa cm^3/g and a specific modulus of 0.69 ± 0.03 GPa cm^3/g and total strain of 6.90 ± 0.53%.

D. L. S. Rosa · M. O. de Lima Júnior · A. C. R. da Silva · V. S. Candido (✉)
Materials Science and Engineering Program, Federal University of Para—UFPA, Ananindeua, Brazil
e-mail: scarpini@ufpa.br

D. L. S. Rosa
e-mail: liradimilira@gmail.com

M. O. de Lima Júnior
e-mail: marcellojunior.eng@gmail.com

A. C. R. da Silva
e-mail: alissonrios@ufpa.br

F. P. D. Lopes · N. T. Simonassi
Materials Science Program, North Fluminense State University—UENF, Campos Dos Goytacazes, Brazil
e-mail: perisse@uenf.br

N. T. Simonassi
e-mail: noansimonassi@uenf.br

S. N. Monteiro
Department of Mechanical and Materials Engineering, Military Institute of Engineering, Rio de Janeiro, Brazil
e-mail: sergio.neves@ime.eb.br

H. A. C. Lopera et al. (eds.), *Natural Fibers and Biocomposites 2026*, The Minerals, Metals & Materials Series, https://doi.org/10.1007/978-3-032-13577-3_25

In flexural testing, the strength increased to 56.44 ± 2.76 MPa cm^3/g, the specific modulus of 2.55 ± 0.12 GPa cm^3/g and strain was 2.24 ± 0.17%. Under Izod impact strength increased up to 31.76 ± 2.7 J/m. ANOVA and Tukey tests confirmed the influence of different reinforcement contents. The fractographic analysis by SEM showed a predominantly brittle behavior of the composites. These results indicate the potential of Matamata fiber for use in the structural components of chair seats and backrests.

Keywords Amazonian fibers · Polyester composites · Eco-friendly furniture

Introduction

Due to changes in environmental legislation in many countries, industries have sought to integrate sustainability into the production and delivery of environmentally friendly products. This reinforces the search for more sustainable materials [1, 2]. In this context, composites have been a widely explored solution due to their promising properties, such as high stress resistance combined with their low specific mass [3].

From this perspective, the use of fully synthetic products manufacture of composites demands high consumption of fossil fuels, which leads to a high carbon footprint [4, 5]. Also, due to their origin from non-renewable resources, can they cause significant environmental damage, in addition to posing potential risks to human health, which to result in restrictions on their large-scale use [6–8]. Currently, this problem is driving the field of research in composite materials, which has sought solutions based on studies of new plant fibers as reinforcements compatible with polymer matrices [7, 9, 10].

Nevertheless, despite their advantages, available polymer matrices are limited by strict environmental regulations. Furthermore, their application in composites can also be limiting due to their highly hydrophobic nature. One alternative would be to replace them with fully degradable matrices, but these are still in the research phase, and their application in seating remains limited, as they do not meet the mechanical performance criteria necessary for the safe functioning of the furniture. Another alternative would be to combine chemical treatments to improve hydrophilic/hydrophobic compatibility. However, this approach can increase the cost of the process and generate byproducts during processing, resulting in a larger carbon footprint, which contradicts the goal of reducing impact.

Currently, to reduce the amount of synthetic materials in the production of high-performance engineering fibrous composites, a widely explored alternative has been to investigate different amounts of new plant fibers, combined with different polymer matrices, with the aim of increasing compatibility and the amount of filler material and reducing the amount of synthetic matrices [5, 11, 12].

In this contexto, one advantage is that plant fibers are readily available in the Amazon region. Moreover, Amazonian fibers, possess desirable properties, have high stiffness, and comparable tensile strength and are biodegradability [13, 14]. Amazonian fibers, such as ubim [5], tucum [15], embira [12], guaruma [14], and titica [4], have gained increasing popularity due to their inherent properties, notably their low density combined with their high strength. A good example is the periquiteira fiber reaching a significant tensile strength value of 178 MPa e density of 0.43 g/cm^3 [16]. Therefore, these fibers have recently demonstrated their potential use as filler materials compatible with polymer matrices, highlighting their potential applications in composite engineering for high-performance products. Thus, the use those material has become increasingly popular in sectors such as automotive, construction, and ballistic armor [17].

Among these available fibers, Matamata fiber emerges as a new candidate in the furniture area, standing out as a viable alternative for its application as compatible reinforcements in polyester matrix for incorporation into furniture components, such as chair seats and backrests. The matamata fiber has a density of 0.38 g/cm^3, specific strength of 152.07 MPa cm^3/g, crystallinity index of 63.54% and low microfibrillar angle of 6.12°, in addition to having high thermal resistance up to 486 °C [18].

Therefore, this article aims to evaluate the mechanical behavior of composites manufactured by hand molding with 10, 20, and 30 vol% Matamata fibers and polyester resin as a matrix, tested primarily by tensile, flexural, and Izod impact, as well as examined using by SEM. Thus, this article is relevant because it points to new paths for the improvement of polymeric composite materials, considering the matamatá fiber, a new Amazonian fibers for specific use in ecological furniture, aiming at safer structural applications in interior furnishings. So, this work is relevant for the development of new products to be manufactured from new fibers amazonian, aiming at improving their use in engineering composites.

Methodology

Materials

Fibers from the bark of the matamata tree (purchased from local producers), unsaturated terephthalic polyester resin (CENTERPOL C400/POLY 500) pre-accelerated with 0.2% cobalt naphthalene (CoNap), and 2 vol% methyl ethyl ketone peroxide (Butonax M – 50) as a hardener. All products were supplied by ARAQUIMICA, Brazil.

Manufacturing of Composites

The process adopted lay-up molding, in which the fibers were oriented longitudinally within silicone molds, according to Fig. 1. Polyester resin was used in combination

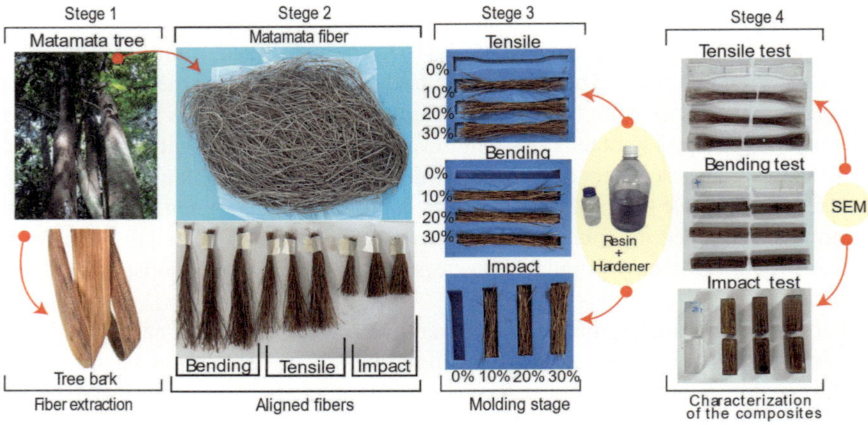

Fig. 1 Key processing steps for molding composites with 0, 10, 20, and 30 vol% matamata fiber

Table 1 Investigated compositions

Samples	Fiber (vol%)	Resin (vol%)	Hardener (vol%)	Number of specimens
CP 0	0	100	2.0	8
CP 10	10	90	2.0	8
CP 20	20	80	2.0	8
CP 30	30	70	2.0	8

with natural matamata fibers, added in controlled volumes of 10, 20 and, 30 vol% as specified in Table 1.

In step 1, the bark was extracted from the tree, washed in running water, and dried under ambient conditions. In step 2, the fibers were separated into controlled volumes, and aligned longitudinally into molds. In step 3, the fibers were placed in silicone molds and poured with polyester resin homogenized with 2 vol% hardener. In step 4, the samples were tested for tensile, flexural, and Izod impact strength, and characterized by SEM.

Mechanical Characterization of Composites

Density of the Composites

ASTM D 2734 [19], was used to determine the density of 6 samples (30 × 12.40 × 12.40 mm), prepared with contents of 10, 20, and 30 vol% of aligned matamatá fibers and polyester resin with 2% hardener. After curing, the samples were dried at 100 °C for 20 min, measured, and weighed on an analytical balance (± 0.0001 g) in triplicate. The density was calculated by Eq. (1):

$$\rho_{CP} = \frac{m}{v} \tag{1}$$

where: ρ_{CP} is the density of the composite in (g/cm^3), m is the mass in (g), v is the volume in (cm^3).

The efficiencies of the composites was be calculated using Eq. 2, allowing the determination of the structural efficiency of each composite added with 10, 20 and 30 vol% of fibers.

$$\eta = \frac{\mathrm{E_{Cp}}}{\mathrm{C}} \tag{2}$$

Being: η the efficiency of the composite, E_{CP} is the strength property of the composite and C is the density of the composite.

Tensile Test

ASTM D 638 [20] was used to prepare 8 test specimens (162 × 20 × 3.2 mm), tested on a universal electromechanical machine (INTERMÉTRIC iM50, Brazil), using a 5 kN load cell and a speed of 0.5 mm/min.

Bending Test

ASTM D790 [21] was used to manufacture 6 rectangular specimens (127 × 12.7 × 6.4 mm) that were subjected to testing in a universal electromechanical machine (INTERMÉTRIC iM50, Brazil), with a 5 kN load cell, a distance between supports of 60 mm and a speed of 2 mm/min.

Izod Impact Test

ASTM D256 [22] was used to prepare 6 prismatic specimens (63.5 × 12.7 × 12.7 mm) with a 2.54 mm notch. The tests were conducted on a Tinius Olsen impact machine (model IT542), equipped with a PANTEC instrumented pendulum (model XC-50, 200 V, 60 Hz).

Statistical Analysis

The results were evaluated by analysis of variance (ANOVA) with 95% confidence ($p < 0.05$). When $p < 0.05$ was identified, the Tukey's test was applied to compare the means.

Scanning Electron Microscopy (SEM)

Samples of CPs at 0, 10, 20, and 30 vol%, obtained after mechanical tests, were used. All samples were metallized with gold (Au) and analyzed using a Mira 3 FEG 250 TESCAN microscope (Brno, Czech Republic), operating with secondary electrons at 5–15 kV, working distance of 10–25 mm, and magnifications of 90–1000 × .

Results and Discussion

Density of Composites

The densities of each composite expressed in g/cm^3 can be found in Table 2.

According to these results, the composites with 10, 20 and 30 vol% of fiber showed densities of 1.21 ± 0.060, 1.18 ± 0.011, 1.17 ± 0.011 g/cm^3, respectively. Despite these marginals densities results, a greater contribution was recorded for the composite with 10 vol% fiber, followed by 20 and 30 vol%, respectively. This can be explained by the low density contribution conferred by the Matamata fiber (0.38 g/cm^3), compared to the contribution of the resin with 1.15 g/cm^3. This implies that the resin contributes the greatest density for the lower compositions, which explains the lower densities found for 20 and 30 vol%, since a smaller amount of resin will occupy a smaller filler area in the structure of these composites, which corresponds to greater contributions from the fiber as filler in the higher compositions, impacting in lower densities, as noted for the composite with 20 and 30 vol% fiber compared to 10 vol%.

Tensile Test

Figure 2 presents the specific resistance results obtained under uniaxial tension, where in (a) are the stress/strain curves, in (b) the specific tensile resistance, and in (c) the specific modulus and the strain under tension.

After tensile testing, the CPs with 0,10, 20, and 30 vol% of matamata fibers exhibited specific strengths of 30.80 ± 1.73, 31.62 ± 0.87, 36.41 ± 1.61 and

Table 2 Results of the density test of composites with 10, 20, and 30 vol% of fiber

Composite (%vol)	Density (g/cm^3)
10	1.21 ± 0.060
20	1.18 ± 0.011
30	1.17 ± 0.011

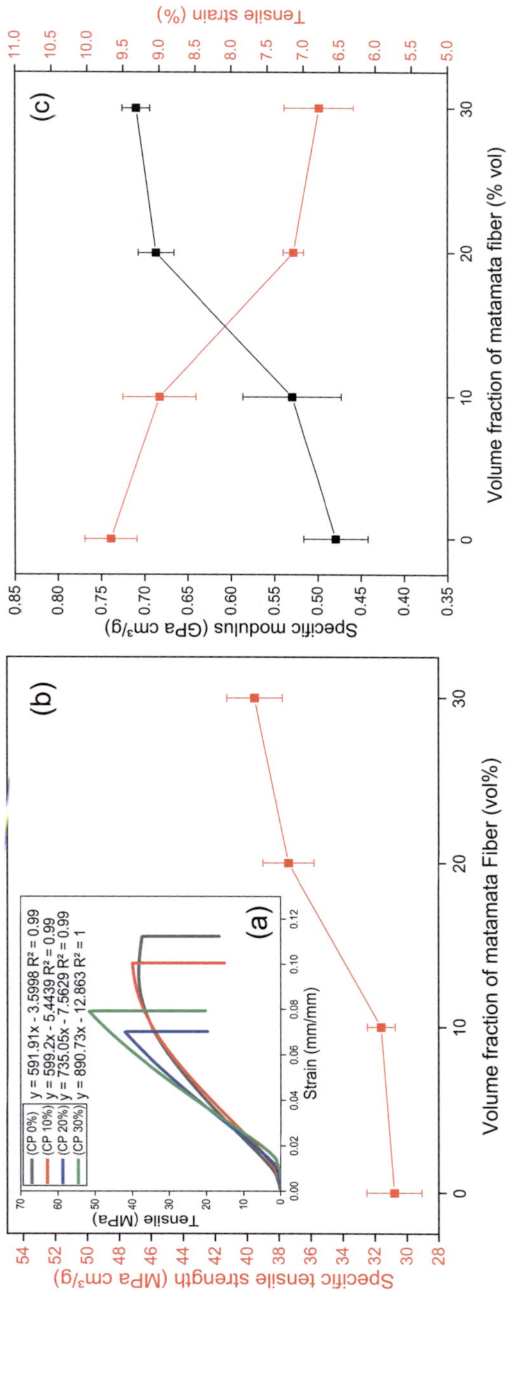

Fig. 2 Results of the mechanical properties of the specimens (CPs) obtained using a 5 kN load cell and a displacement rate of 0.5 mm/min

39.51 ± 2.60 MPa cm^3/g, respectively. The percentage increases were 2.66% (0/10%), 18.21% (10/20%) and 28.28% (20/30%). A progressive increase in strength of approximately 28% compared to the isolated matrix is observed, demonstrating that fibrous reinforcement plays a decisive role in supporting applied stresses. In practical terms, this improvement implies greater load-bearing capacity and mechanical stresses, an essential requirement for structural applications in furniture, where chair seats are subjected to concentrated and repetitive loads.

Likewise, the specific module showed considerable improvements with the variation of 0,10, 20, and 30 vol% of matamata fibers, starting with 0.48 ± 0.04 GPa cm^3/g to 0.53 ± 0.06 GPa cm^3/g, reaches 0.68 ± 0.02 and 0.72 ± 0.02 GPa.cm^3/g in the largest fractions, respectively. Consequently, the percentage increases were 10.42% (0/10%), 41.67% (10/20%) and 50% (20/30%). That is, the specific modulus increases from 0.48 (0%) to 0.72 GPa.cm^3/g (30%), or approximately a 50% improvement in the relative stiffness of the composite. This increase provides greater dimensional stability and a lower tendency to deform under static load, favoring use in seats that need to withstand stress without presenting permanent deformation.

In terms of the deformation, the contents of 10, 20 and 30%vol showed a progressive decline in relation to the isolated matrix that reached9.67 ± 0.36. Thus, the observed percentages were 8.99 ± 0.51, 7.14 ± 0.14 and 6.78 ± 0.48%, presenting an inverse trend to the specific module with a reduction from 9.67% (0%) to 6.78% (30%). The reductions in observed deformation were 7.03% (0/10%), 26.16% (10/20%) and 29.89% (20/30%). Specifically in fibrous composite materials, this loss of ductility is characteristic of the insertion of fibers into the matrix, making the material more rigid, as reported in previous research on similar materials [23–29]. However, for bench and chair seats, a moderate level of deformation is desirable, as it prevents sudden failures from propagating through the matrix, ensuring greater user safety. Thus, the results of the tensile test validated by ANOVA can be found in Table 3.

The ANOVA analysis confirmed significant differences between the means of the specific tensile strength properties, where the values of F resulted in 28.112, 48.746, 61.695, respectively, much larger than the calculated critical F values all equal to 3.098, corroborating that these differences are influenced by the modulation in the formulation and not originating from external fluctuations. Therefore, since the differences were reportedly supported by ANOVA. The next step was to apply Tukey's test to qualify the detected differences. Therefore, the results of the Tukey test are shown in Table 4.

Tukey's test showed significant differences between the samples, exceeding the msd calculated equal to 3.20, 0.07, 0.78, respectively.The main differences were detected between the pairs 0/20%, 0/30%, 10/20%, 10/30%, respectively, with maximum differences of 6.6, 8.7, 5.8, 7.9 (Specific tensile strength), 0.21, 0.24, 0.16, 0.19 (Specific Modulus strength) and 2.53, 2.89, 1.86, 2.21 (Flexural Strain), respectively, indicating that the differences arise from changes in the formulation of the samples due to the addition of matamata fibers. Considering the balance between strength, stiffness, and deformation, the composite with 20 vol% fibrous reinforcement proved to be more promising, as it provides increased specific strength

Table 3 Results of the ANOVA for the specific properties strength: tensile strength, elastic modulus, and tensile strain

Source	SQ	gl	MQ	F	P-value	F critical
Specific tensile strength (MPa cm³/g)						
Between groups	330.81	3	110.27	28.112	2.25×10^{-07}	3.098
Inside the group	78.45	20	3.92			
Total	409.26	23				
Specific modulus (GPa cm³/g)						
Between groups	0.2469	3	0.0823	48.746	2.22×10^{-09}	3.098
Inside the group	0.0338	20	0.0017			
Total	0.2807	23				
Tensile strain (%)						
Between groups	35.493	3	11.831	61.695	2.75×10^{-10}	3.098
Inside the group	3.835	20	0.192			
Total	39.328	23				

Table 4 Results of the Tukey's test for specific tensile strength properties

Composition	0%	10%	20%	30%
Specific tensile strength (msd = 3.20)				
0%	–	0.8	**6.6**	**8.7**
10%	0.8	–	**5.8**	**7.9**
20%	**6.6**	**5.8**	–	2.1
30%	**8.7**	**7.9**	2.1	–
Specific modulus strength (msd = 0.07)				
0%	–	0.05	**0.21**	**0.24**
10%	0.05	–	**0.16**	**0.19**
20%	**0.21**	**0.16**	–	0.03
30%	**0.24**	**0.19**	0.03	–
Tensile strain (msd = 0.78)				
0%	–	0.68	**2.53**	**2.89**
10%	0.68	–	**1.86**	**2.21**
20%	**2.53**	**1.86**	–	0.35
30%	**2.89**	**2.21**	0.35	–

(36.41 MPa cm^3/g) and specific modulus (0.68 GPa cm^3/g), while maintaining deformation at 7.14%, still sufficient to ensure energy absorption and reduce the risk of sudden failure, while a 30 vol% content prioritizes mechanical performance at the expense of ductility. Therefore, the 20 vol% composition achieves the most appropriate balance between strength and toughness.

Bending Test

Figure 3 shows the results of the mechanical properties of the composites obtained under bending, where, in (a) are the stress–strain curves, in (b) the specific flexural strength, and in (c), the specific modulus of elasticity versus the deformation, all depending on the composition of the samples.

Under flexion, the composites showed a progressive increase in specific flexural strength with contents of 10, 20, and 30 vol%, resulting in 53.53 ± 5.20, 54.25 ± 5.62, 58.18 ± 4.54 MPa cm^3/g in relation to the polyester matrix which resulted in 50.16 ± 4.54, with the increases being 6.71% (0/10%), 8.16% (10/20%) and 15.99% (20/30%), respectively. The biggest the observed variation from 50.16 ± 4.54 (0%) to 58.8 ± 4.54 MPa cm^3/g (30%) indicates that composites with higher fiber contents have a greater capacity to resist deflective stresses, with emphasis on 30 vol%, which presented the best performance in this property.

In relation to the specific modulus under bending, the results showed less linear values, going from 2.39 ± 0.12 in 0%, it decays to 2.12 ± 0.05 in 10%, increases to 2.36 ± 0.10 in 20% and reaches 2.66 ± 0.04 in 30%vol, being that the most significant increase in stiffness noted between (20/30%) was 11.30%. However, the highest value of 2.66 GPa.cm^3/g, results in a stiffer composite, less prone to deformation when subjected to bending, as this parameter is directly associated with the material's stiffness. Specifically for seats, stiffness is positive to a certain extent, as it ensures dimensional stability. However, excessive stiffness can lead to sudden catastrophic failures, compromising user safety. Thus, composites with specific modulus around 2.3 GPa cm^3/g should be considered, as they can offer a more appropriate balance between stability and flexibility.

The deformation also resulted in slightly linear values, varying progressively with 0, 10, 20, and 30 vol%, with the recorded values being equal to 2.39 ± 0.12%, 2.12 ± 0.05%, 2.36 ± 0.10% and 2.66 ± 0.04%, respectively. Thus, the most significant deformation increments were 19.63% (0/10%), 19.63% (10/20%), and 14.61% (20/30%). Since this property is a crucial indicator for evaluating toughness and pre-rupture behavior, intermediate values (≈ 2.51%) indicate that these composites can absorb larger deformations, which is crucial for furniture subjected to repetitive stress (i.e., sitting and standing movements). Thus, composites within the range deformations (2.51–2.62%) can offer greater safety against catastrophic failures.

In seating, the balance between strength, modulus, and deformation ensures that less material is required to achieve adequate mechanical performance, reducing fiber extraction and extending the product's useful life. These aspects converge with current demands for sustainability in materials engineering, which is in line with previous studies [30–34]. Thus, the results of the ANOVA validated flexion test can be found in Table 5.

The ANOVA analysis confirmed significant differences between the means of the specific flexural strength properties, where the values of F resulted in 28,586 (Specific Flexural Modulus) and 9.761 (Flexural Strain), far exceeding the values of F-critical equal to 3.24, corroborating the rejection of the null hypothesis. Therefore,

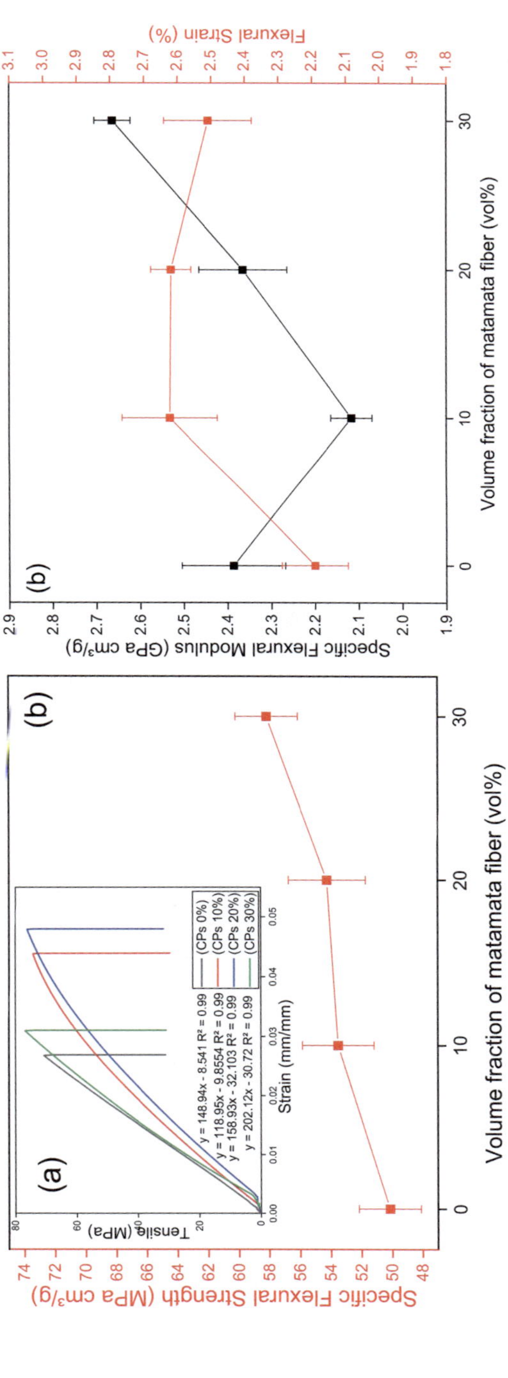

Fig. 3 Results of the mechanical properties of the specimens obtained under biaxial bending, using a 5 kN load cell and a displacement rate of 2 mm/min

Table 5 Results of the ANOVA for the mechanical properties specifics: tensile strength, elastic modulus, and flexural strain

Source	SQ	gl	MQ	F	P-volume	F-critical
Specific flexural strength (MPa cm^3/g)						
Between groups	162.48	3	54.16	1.73	0.200	3.24
Inside the group	499.54	16	31.22			
Total	662.030	19				
Specific flexural modulus (GPa cm^3/g)						
Between groups	0.751	3	0.2503	28.586	1.16 × 10–06	3.24
Inside the group	0.140	16	0.0088			
Total	0.891	19				
Flexural strain (%)						
Between groups	0.522	3	0.1739	9.761	0.000671	3.24
Inside the group	0.285	16	0.0178			
Total	0.807	19				

the stated differences were clearly supported by ANOVA. For the specific flexural strength, no margin of differences was detected. As a next step, Tukey's test was applied to the detected differences. The Tukey's test conducted from the bending test can be found in Table 6.

Tukey's test showed significant differences between the samples, exceeding the calculated msd equal to 0.140 (Flexural Specific Modulus) and 0.200 (Flexural Strain). The main differences were detected in the pairs 0/10%, 0/30%, 10/20%, 10/30%, 20/30%, with maximum differences of 0.270, 0.277, 0.249, 0.547 and 0.299 (Specific Flexural Modulus) and 0.431, 0.257, 0.318 and 0.257 (Flexural Strain), respectively. Therefore, from a flexural test perspective, the 30 vol% formulation stands out for its combination of greater strength and stiffness (58.18 MPa cm^3/g,

Table 6 Results of the Tukey test for specific flexural strength properties

Composition	0%	10%	20%	30%
Specific flexural modulus (msd = 0.140)				
0%	–	**0.270**	0.021	**0.277**
10%	**0.270**	–	**0.249**	**0.547**
20%	0.021	**0.249**	–	**0.299**
30%	**0.277**	**0.547**	**0.299**	–
Flexural strain (msd = 0.200)				
0%	–	**0.431**	0.174	**0.318**
10%	**0.431**	–	**0.257**	0.113
20%	0.174	**0.257**	–	0.144
30%	**0.318**	0.113	0.144	–

2.66 GPa cm^3/g) while maintaining deformation in (2.51%), making it a promising option for chair seats requiring greater durability and mechanical support. However, the 20 vol% fiber formulation, on the other hand, presents a balance between strength, moderate stiffness, and greater deformation (54.25 MPa cm^3/g, 2.36 GPa cm^3/g, and 2.62%), which may be ideal in contexts where comfort and fatigue resistance are priorities. The 10 vol% formulation, exhibited lower stiffness and a higher observed deformation rate (53.53 MPa cm^3/g, 2.12 GPa cm^3/g and 2.62%), making it a good option in contexts that require more pronounced deformations.

Izod Impact Test

Figure 4 presents the results of the Izod impact test, with (a) showing the energy absorption configuration and (b) the efficiency of composites.

The average Izod impact energy values were progressively higher with increasing reinforcement content in the polyester matrix, with the values recorded for samples with 0, 10, 20 and 30 vol% fibers have been 11.17 ± 0.72 J/m, 17.03 ± 0.65 J/m, 21.35 ± 1.84 J/m and 31.76 ± 2.7 J/m, respectively. Consequently, the percentage increases were 52.46% (0/10%), 25.37% (10/20%), 48.76% (20/30%), being a significant increase of 184.33% (0/30%) in relation to the polyester matrix, evidencing an increase in the fibrous reinforcement efficiency, directly improving this property. Previous research also reported increases in impact energy with the increase in fiber content in the matrix, such as for carnauba fibers [32], tucum [15], ubim [5], among others [35–38].

The efficiencies for composites with 0, 10, 20, and 30 vol% by volume fiber were 0.01, 0.014, 0.018, and 0.027, respectively. These values are marginal to each other, however, they indicate that adding greater amounts of matamata fiber gradually improves the composite's efficiency when considering impact energy. Therefore, from an impact energy perspective, the higher compositions (20 and 30%) are preferable to compositions with lower fiber contents. Thus, the results of the ANOVA conducted for the impact energy are expressed in Table 7.

ANOVA confirmed significant differences between the averages of Impact energy, where the value of F equal to 167.54, it far exceeded the value of F-critical equal to 3.01, expressing a value of P-volume equal to 3.17×10^{-16}, much lower than the significance level of 0.05, corroborating the rejection of the null hypothesis. Therefore, the differences detected were clearly supported by ANOVA; it remains to be seen in which groups of means these differences are predominant. For this, Tukey's test was applied. The Tukey's test conducted for the same properties from the impact test can be found in Table 8.

Using the Tukey's test, significant differences were attributed between the samples, in such a way that it exceeded the calculated msd with a value of 3.14. The main differences were detected in the pairs 0/10%, 0/20%, 0/30%, 10/30%, 20/30%, with maximum differences of 5.856, 10.176, 20.593, 14.738, and 10.417, respectively, reflecting the dominant influence of reinforcement fibrous content on impact

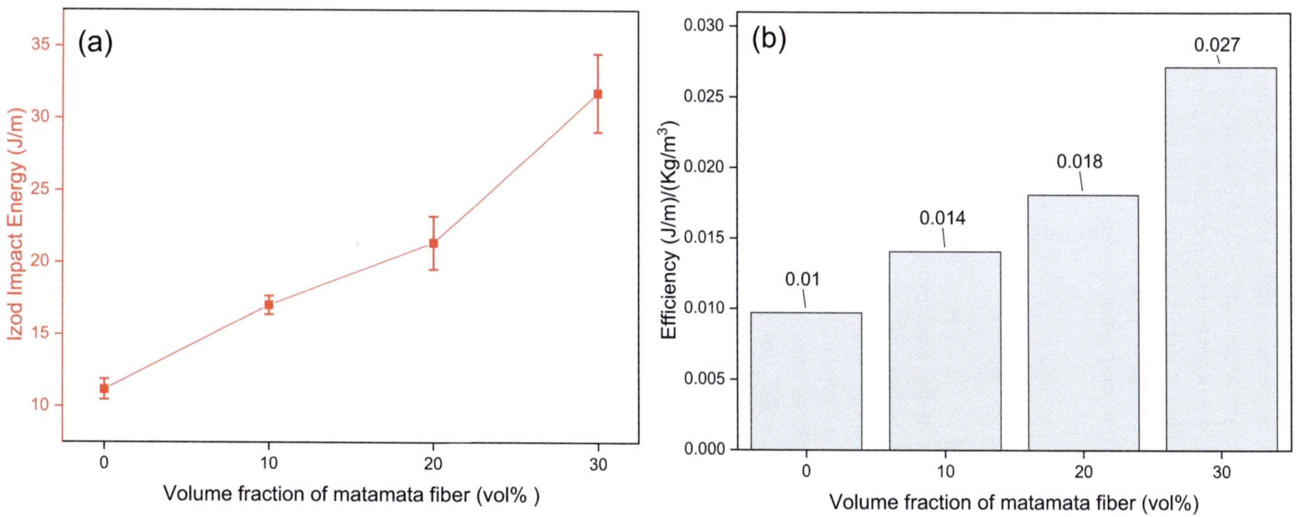

Fig. 4 Results of the Izod impact test. In (**a**), the impact energy is shown, and in (**b**) the efficiency of composites

Table 7 ANOVA results for IZOD impact resistance properties

Source	SQ	gl	MQ	F	P-volume	F-critical
Between groups	1586.06	3	528.68	167.54	3.17×10^{-16}	3.01
Inside the group	75.73	24	3.16			
Total	1661.79	27				

Table 8 Tukey's test results for impact resistance properties

Composition	0%	10%	20%	30%
Izod impact energy (msd = 3.14)				
0%	0	**5.856**	**10.176**	**20.593**
10%	**5.856**	0	4.320	**14.738**
20%	**10.176**	4.320	0	**10.417**
30%	**20.593**	**14.738**	**10.417**	0

energy. That is, the observed differences in impact energy are primarily attributed to formulation modulation. Therefore, from the perspective of impact energy absorption, the 30 vol% fiber sample is more suitable for use in furniture seating, as it exhibits higher average energy absorption. However, it is important to highlight that the 20 vol% fiber formulation optimizes impact energy by using a smaller amount of fiber, can making it more suitable for sustainable applications.

SEM from the CPs Tensile and Bending Tests

Figure 5 presents the fractographic analysis by SEM, where the morphology of the cross-section of the fractured samples after the tensile and flexural test is displayed, being (a), (b), (c) and (d) the CPs with 0, 10, 20 and 30 vol% of fibers.

Fractographic analysis showed that, in both tensile and flexural stresses, all composites exhibited predominantly brittle behavior, characterized by the presence of river marks and crack propagation typical of brittle polymer matrices. CP 0% exhibited delamination and greater porosity in flexural stress, resulting from the manual method, while CP 10% exhibited low homogeneity in fiber dispersion, both in tensile and flexural stresses, as well as delamination and interfacial fracture, corroborated by the presence of crack propagation in the polyester matrix structure. CP 20% exhibited better fiber dispersion, both in tensile and flexural stresses, with little evidence of point fiber pullout, suggesting good interfacial adhesion, demonstrating greater absorption and redistribution of stresses within the matrix, indicating partial action as a physical barrier. At CP 30%, the fracture maintained its brittle nature, but revealed greater dispersion in both tensile and flexural, and less fiber agglomeration in tensile than in

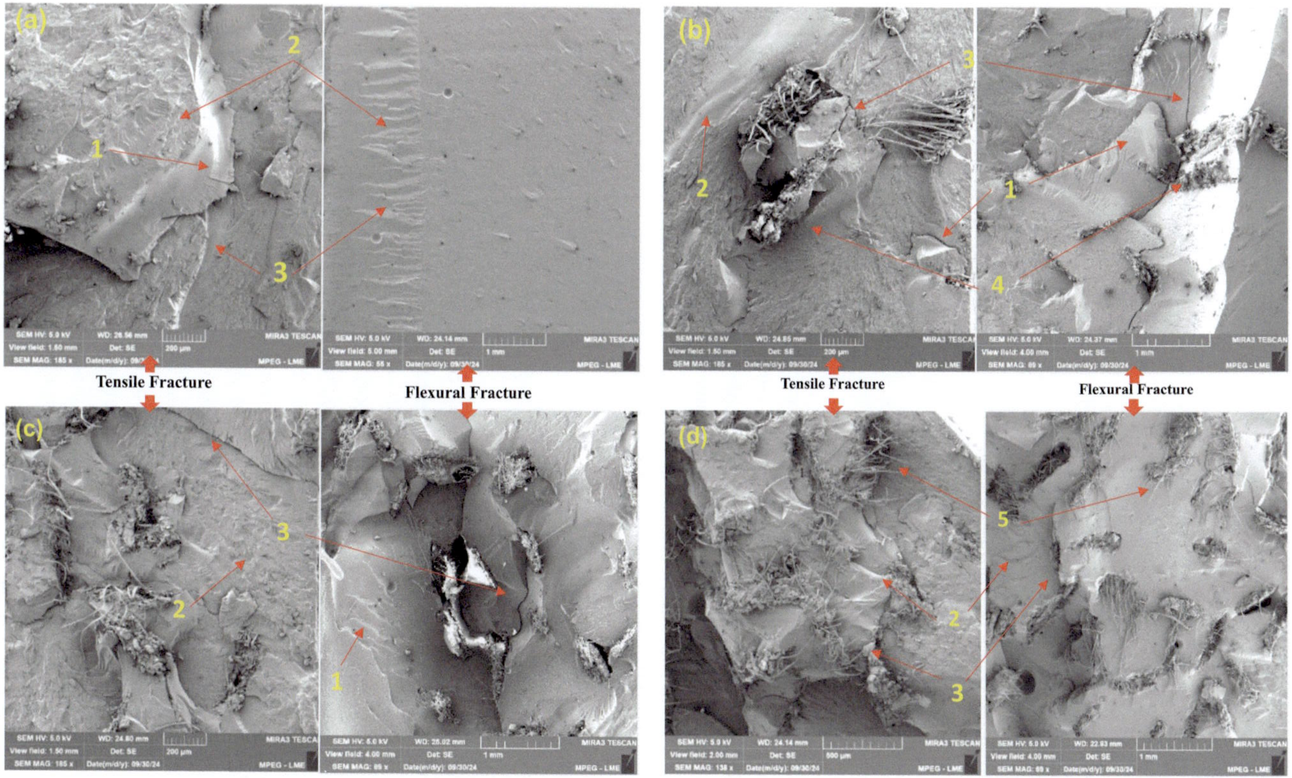

Fig. 5 Micrographs of the composites, where: (**a**), (**b**), (**c**) and (**d**) are the CPs with 0, 10, 20 and 30 vol% after the tensile and flexural tests, respectively, and where: 1. Delamination, 2. River marks, 3. Crack propagation, 4. Fiber crack, 5. Protrusion

flexural, with few cohesive failures at the fiber/matrix interface, signs of interfacial fracture and fibril protrusion, which indicates more effective reinforcement.

In general, the 20 and 30 vol% contents demonstrated greater efficiency in reinforcement, although the behavior of the composites is predominantly fragile, which is in line with previous studies on fibrous composites [26, 33, 35, 37, 38]. However, the propagation of cracks in the matrix towards the fibers suggests that they reinforce the composites more effectively, especially by absorbing and redistributing internal stresses, mainly in composites with 20 and 30 vol% of fibers, a behavior that has also been described in previous works [28–31, 35, 37].

Conclusion

In tensile test, an increase in specific resistance from 31.87 ± 1.62 MPa cm^3/g (0%) to 39.08 ± 2.62 MPa cm^3/g (30%) was observed, corroborated by the variation in specific modulus, which varied from 0.52 ± 0.02 GPa cm^3/g (0%) to 0.69 ± 0.03 GPa cm^3/g (30%), and by the greater variation in driven deformation, from 9.44 ± 0.20% to 6.90 ± 0.53%. In bending test, the resistance increased from 51.09 ± 5.42 MPa cm^3/g (0%) to 56.44 ± 2.76 MPa cm^3/g (30%), while the elastic modulus showed less linearity, varying from 2.44 ± 0.12 GPa cm^3/g to 2.55 ± 0.12 GPa cm^3/g, and the deformation was less variable, starting at 2.10 ± 0.25% (0% vol.) and reaching 2.24 ± 0.17% (30% vol). Under izod impact, the results were more expressive: it was found that the impact energy increased significantly with the fiber content, going from 11.17 ± 0.72 J/m (10%) to 31.76 ± 2.7 J/m (30%). The specific efficiency also increased with the fiber content, going from 0.010 to 0.027.

Therefore, the formulation with 20% fiber content, with 58.18 MPa cm^3/g flexural strength, 2.66 GPa cm^3/g modulus and 2.51% deformation, is the most suitable for chair seats, combining high structural performance, comfort and lower environmental impact. This composition can balance mechanical performance and structural safety, requirements that are fundamental for furniture that must be both strong and durable.

References

1. Almtori SAS, Jassim RJ, Ali DC, Kadhim EH, Al-Sabur R (2023) Sustainable manufacturing process applied to produce waste polymer-polymer matrix composites. Environ Res Eng Manag 79(1):122–132. https://doi.org/10.5755/j01.erem.79.1.32907
2. Nurazzi NM, Asyraf MRM, Khalina A, Abdullah N, Aisyah HA, Rafiqah SA et al (2021) A review on natural fiber reinforced polymer composite for bullet proof and ballistic applications. Polymers 13(4):646. https://doi.org/10.3390/polym13040646
3. Santos AGA, Nascimento D, da CB, Lopes FPD, Simonassi NT, Monteiro SN, Silva ACR da et al (2025) Investigation of the flexural and charpy impact properties of polymer composites reinforced with Tururi (Manicaria saccifera) fibrous fabric. Polymers 17(4):466. https://doi.org/10.3390/polym17040466

4. Cunha JDSCD, Nascimento LFC, Luz FSD, Garcia Filho FDC, Oliveira MS, Monteiro SN (2022) Titica vine fiber (Heteropsis flexuosa): a hidden amazon fiber with potential applications as reinforcement in polymer matrix composites. J Compos Sci. 6(9):251. https://doi.org/10.3390/jcs6090251
5. Marchi BZ, Oliveira MS, Bezerra WBA, de Sousa TG, Candido VS, da Silva ACR et al (2022) Ubim Fiber (Geonoma baculífera): a less known Brazilian amazon natural fiber for engineering applications. Sustainability 14(1):421. https://doi.org/10.3390/su14010421
6. Antony S, Cherouat A, Montay G (2021) Effect of fiber content on the mechanical properties of hemp fiber woven fabrics/polypropylene composite laminates. Polym Polym Compos 29(9_suppl):S790–S802. https://doi.org/10.1177/09673911211023942
7. Balogun OA, Daramola OO, Adediran AA, Akinwande AA, Bello OS (2023) Investigation of Jute/Tetracarpidium conophorum reinforced polypropylene composites for automobile application: Mechanical, wear and flow properties. Alex Eng J 65(1):327–341. https://doi.org/10.1016/j.aej.2022.10.026
8. Zhao X, Zhang Z, Pang J, Su L (2023) Preparation of carbon fiber-reinforced composite panels from epoxy resin matrix of nano lignin polyol particles. J Clean Prod 428(1):139170. https://doi.org/10.1016/j.jclepro.2023.139170
9. Adugna Ayalew A, Fenta Wodag A (2022) Characterization of Chemically Treated Sisal Fiber/Polyester Composites. J Eng 1:8583213. https://doi.org/10.1155/2022/8583213
10. Sathyamoorthy G, Vijay R, Lenin Singaravelu D (2021) Development and characterization of alkali-treated and untreated Dactyloctenium aegyptium fibers based epoxy composites. Mater Today Proc. 39(4):1215–1220. https://doi.org/10.1016/j.matpr.2020.03.796
11. Szadkowski B, Śliwka-Kaszyńska M, Marzec A (2025) Improving compatibility between coffee or black tea ground wastes and polymer matrix via silane treatment for producing sustainable biofillers. Sci Rep 15(1):13554. https://doi.org/10.1038/s41598-025-97980-7
12. Tavares SS, de Mendonça NL, Lopera HC, Monteiro SN, Meyers MA (2025) The embira bark fiber: a sustainable Amazon tape. Adv Compos Hybrid Mater. 8(1):184. https://doi.org/10.1007/s42114-024-01170-4
13. Araújo SS, Santos GTA, Tolosa GR, Hiranobe CT, Budemberg ER, Cabrera FC et al (2023) Acai residue as an ecologic filler to reinforcement of natural rubber biocomposites. Mater Res 26 (suppl. 1). https://doi.org/10.1590/1980-5373-MR-2022-0505
14. Lima TESD, Natalli JF, Marvila MT, Candido VS, Silva ACRD, Azevedo ARGD (2024) Potentiality of using guarumã natural fiber (Ischinosiphon koern) in sustainable coating mortars. J Mater Res Technol. 30(1):1301–1309. https://doi.org/10.1016/j.jmrt.2024.03.101
15. Oliveira M, Luz F, Teixeira Souza A, Demosthenes L, Pereira A, Filho F et al (2020) Tucum Fiber from Amazonian Astrocaryum vulgare Palm Tree: novel reinforcement for polymer composites. Polymers 12(10):2259. https://doi.org/10.3390/polym12102259
16. Pinheiro MA, Ribeiro MM, Rosa DLS, Nascimento DCB, Silva ACR, Reis MAL et al (2023) Periquiteira (*Cochlospermum orinocense*): a promising Amazon fiber for application in composite materials. Polymer 15(9):2120. https://doi.org/10.3390/polym15092120
17. Marchi BZ, da Silveira PHPM, Bezerra WBA, Nascimento LFC, Lopes FPD, Candido VS et al (2023) Ballistic performance, thermal and chemical characterization of Ubim Fiber (Geonoma baculifera) Reinforced Epoxy Matrix Composites. Polymers 15(15):3220. https://doi.org/10.3390/polym15153220
18. Santa Rosa DL, Nascimento DDCB, Lopes FPD, Simonassi NT, Monteiro SN, da Silva ACR, Candido VS (2025) Matamata (Lecythis idatimon): a novel lighter cost-effective amazonian fiber for eco-friendly composite applications in furniture design. J Mater Res Technol. 38(1):4027–4048. https://doi.org/10.1016/j.jmrt.2025.08.221
19. American Society For Testing Materials D 2734 (2023) Standard Test Methods for Void Content of Reinforced Plastic. ASTM International: West Conshohocken, PA, USA
20. American Society For Testing Materials D 638 (2023) Standard Test Method for Tensile Properties of Plastics, Annual Book of ASTM Standards, American Society for Testing and Materials

21. American Society For Testing Materials D 790 (2017) Flexural Properties of Unreinforced and Reinforced Plastics and Electrical Insulating Materials, Annual Book of ASTM Standards, American Society for Testing and Materials
22. American Society For Testing Materials D 256 (2023) Standard Test Method for Determining the Izod Pendulum Impact Resistence of Plastics. ASTM International: West Conshohocken, PA, USA
23. Costa UO, Nascimento LFC, Garcia JM, Bezerra WBA, Monteiro SN (2020) Evaluation of Izod impact and bend properties of epoxy composites reinforced with mallow fibers. J Mater Res Technol. 9(1):373–382. https://doi.org/10.1016/j.jmrt.2019.10.066
24. Wondmagegnehu BT, Paramasivam V, Selvaraj SK (2021) Fabricated and analyzed the mechanical properties of textile waste/glass fiber hybrid composite material. Mater Today Proc. 46(17):7297–7303. https://doi.org/10.1016/j.matpr.2020.12.984
25. Almeida LM, da Silva ACR, Lopes FPD, Simonassi NT, Monteiro SN, Candido VS (2025) Caranan fibers (Mauritiella armata) and new reinforcements of polyester composites with natural fibers functionalized with graphene oxide and their application potential. J Mater Res Technol. 34(1):2232–2252. https://doi.org/10.1016/j.jmrt.2024.12.233
26. Vinod A, Vijay R, Singaravelu DL, Elayaperumal A, Sanjay MR, Siengchin S et al (2022) Effect of alkali treatment on performance characterization of Ziziphus mauritiana fiber and its epoxy composites. J Ind Textil 51 (2_suppl):2444S-2466S. https://doi.org/10.1177/1528083720942614
27. Neves ACC, Rohen LA, Mantovani DP, Carvalho JPRG, Vieira MF, Lopes FPD et al (2020) Comparative mechanical properties between biocomposites of epoxy and polyester matrices reinforced by hemp fiber. J Mater Res Technol 9(2):1296–1304. https://doi.org/10.1016/j.jmrt.2019.11.056
28. Kumar SS, Vignesh V, Prasad VVSH, Rajeshkumar L, Karthikeyan S, Uthayakumar M (2024) Static and dynamic mechanical analysis of natural fiber hybrid composites for engineering applications. Biomass Convers Biorefinery 14:14889–14901. https://doi.org/10.1007/s13399-022-03689-2
29. Gashawtena E, Kidane A, Sirahbizu B (2024) The effect of nanocellulose and silica filler on the mechanical properties of natural fiber polymer matrix composites. Results Eng. 24(1):102898. https://doi.org/10.1016/j.rineng.2024.102898
30. Pawar US, Chavan SS, Mohite DD (2024) Synthesis of glass FRP-natural fiber hybrid composites (NFHC) and its mechanical characterization. Discover Sustainability 5(44):2024. https://doi.org/10.1007/s43621-024-00231-4
31. Sathees Kumar S (2020) Effect of natural fiber loading on mechanical properties and thermal characteristics of hybrid polyester composites for industrial and construction fields. Fibers Polym 21:1508–1514. https://doi.org/10.1007/s12221-020-9853-4
32. Junio RFP, Nascimento LFC, Neuba LDM, Souza AT, Barbosa JVM, Garcia Filho FDC et al (2020) Copernicia prunifera leaf fiber: a promising new reinforcement for epoxy composites. Polymers 12(9):2090. https://doi.org/10.3390/polym12092090
33. Raja S, Rajesh R, Indran S, Divya D, Suganya Priyadharshini G (2022) Characterization of industrial discarded novel Cymbopogon flexuosus stem fiber: a potential replacement for synthetic fiber. J Ind Textil 51 (1_suppl):1207S-1234S. https://doi.org/10.1177/15280837211007507
34. Tavadi AR, Basheer D, Melese KG, Kumaresan K, Mohan N, Kerur MR et al (2025) Experimental analysis of mechanical properties and FTIR analysis of areca fiber reinforced epoxy composites incorporating Al_2O_3. J Indust Text 55. https://doi.org/10.1177/15280837241313217
35. Pratiwi H, Kusmono Wildan MW (2025) Development and characterization of polyester/ramie fiber hybrid composites reinforced with crystalline nanocellulose extracted from durian peel waste. J Mater Res Technol. 34(1):1201–1212. https://doi.org/10.1016/j.jmrt.2024.12.159
36. Omprakasam S, Vishnu P, Ishanna BM, Sree SA, Rani RD (2025) Banana fiber with madar fiber composites for sustainable printed circuit boards. Mater Lett 380(1):137750. https://doi.org/10.1016/j.matlet.2024.137750

37. Raja T, Raja KV, Reddy MI, Kumar R, Kumar A, Singh V et al (2023) Studies on mechanical and morphological behaviors of banyan/kevlar fibers reinforced MgO particles hybrid aliphatic epoxy composite. Int J Adv Manuf Technol. https://doi.org/10.1007/s00170-023-11852-w
38. Kharchi N, Barka B, Rouabah F, Mourad AHI, Dadache D, Magali A et al (2025) Citric acid treatment: a new approach to improving Ampelodesma mauritanica fibers-poly-lactic acid composites. Ind Crop Prod 223(1):120143. https://doi.org/10.1016/j.indcrop.2024.120143

Vat Photopolymerization Additive Manufacturing of Rice Husk Composites

Omar A. Pulgarín Isaza and L. Henry A. Colorado

Abstract This research investigated composites made of rice husk with photopolymeric resin using the vat photopolymerization additive manufacturing technique. Samples of 0, 2.5, 5.0, and 10wt% were prepared. Several parameters were varied to see their effect on the mechanical response. The tensile tests were conducted in a universal testing machine at 1 mm/min, while their microstructure was observed in an optical and a scanning electron microscopy. Fourier transformed infrared spectroscopy and density tests were conducted as well. Results show that rice husk is feasible to be used as reinforcement for the resin, and based on the amount of this waste in many countries can be a sustainable business for diverse functional applications.

Keywords Additive manufacturing · 3D printing · Rice husk · Vat photopolymerization · Mechanical properties

Introduction

Rice husk (RH) is an abundant and low-cost agro-industrial byproduct that, thanks to its high porosity, low weight, lignocellulosic composition, and significant amorphous silica content, has established itself as a valuable resource in the development of sustainable materials. It is produced in many countries, containing 75–90% of organics (cellulose, lignin, etc. and the rest silica, alkalis, and trace elements [1]. Applications include biomass sources for the extraction of natural silica and then silicon-based products [2], energy generation [3], agriculture [4], and concrete [5]. The variety of applications show is a valuable product due to its composition and mechanical properties. The use of waste materials in upcycling strategies, which

O. A. Pulgarín Isaza
CCComposites Laboratory, Universidad de Antioquia UdeA, Medellín, Colombia

L. H. A. Colorado (✉)
Facultad de Ingeniería, Universidad de Antioquia, Medellín, Colombia
e-mail: henry.colorado@udea.edu.co

H. A. C. Lopera et al. (eds.), *Natural Fibers and Biocomposites 2026*, The Minerals, Metals & Materials Series, https://doi.org/10.1007/978-3-032-13577-3_26

is the creative process of transforming waste materials or unwanted products into new items of higher quality, value, or usefulness, it is always welcome for the both the environmental and economical sustainability of current technologies [6, 7].

In terms of the rice husk as reinforcement of materials, it has been used in cementitious matrices to improve toughness and mitigate cracking, as well as a potential cement reducer due to its pozzolanic activity [8] (when used as ash) and thus to reduce carbon footprint [9]. When used in polymer matrix composites, [10], improves properties such as thermal resistance, compression strength, impact strength, and water absorption, quite important properties in composites. Similarly, Additive Manufacturing (AM) or 3D printing, can also be benefited of the upcycling of this abundant waste as a raw material, giving not only better technical properties, but also contributing to improve the impact of manufacturing in the environment. Thus, rice husk shows great potential for the use in 3D printed parts, already investigated in thermoplastic filament fabrication for Fused Deposition Modeling (FDM) with polylactic acid (PLA) as matrix, with excellent results [11]. The use of natural fibers in AM is a growing field due to its low costs and benefits in mechanical, thermal, and biological properties, although when used in FDM or Direct Ink Writing (DIR), both AM technologies, the surface quality can be limited, [12, 13]. The use of vat photopolymerization, a high-quality surface finishing maker AM technology, can solve this limitation [14–16].

The following research shows results of using vat photopolymerization AM for making polymer matrix composites, using a photocurable resin with rice husk from the industry of Colombia.

Methodology

Rice husks were crushed and subsequently ground in a fiber mill. After this process, the milled product was passed through a series of sieves to ensure a final particle size between 150 and 250 μm. A 3D model of the Type IV tensile specimens was designed according to ASTM D638 so they could be replicated on the Anycubic Photon Basic resin printer with Liquid Crystal Display (LCD) printing technology. The rice husk powders were mixed with Anycubic Basic light-curing resin (skin-colored) in a 5% weight-to-weight ratio to compare the differences in mechanical behavior under tensile stresses.

Three specimens of pure resin and three specimens of resin at a 5% concentration were printed with a 5-min post-curing period. Using the data obtained from the tensile tests, stress–strain graphs are created to analyze the modulus of elasticity and determine the influence of these fibers on the resin matrix. Finally, the structure of the fiber-printed specimens is analyzed in the SEM to identify possible physical alterations.

Results

Figure 1a shows the rice husk as obtained from the rice agricultural industry. Figures 1b and c show Scanning Electron Microscopy (SEM) images of the rice husk after the grinding process showing particles with a distribution between 1 and 20 μm. These small sizes are necessary to reach better mechanical properties due to a better homogeneous distribution of particles and better particle to matrix compatibility, which improves the lead transmission from the matrix to the reinforcement.

Three rice fiber-free resin specimens were printed (see Fig. 2a). Subsequently, after milling, rice husk powders were obtained with the appropriate particle size for proper resin homogenization, which would be used to print three more tensile specimens with a 5% concentration (see Fig. 2b). The surface finishing of samples clearly very smooth and well done.

Following this, stress–strain graphs were obtained (see Fig. 3), which show that adding these powders did not significantly improve the mechanical properties. On the contrary, it showed a slight decrease in the elastic modulus and a lower maximum elongation.

Rice husk is one of the more abundant natural waste fibers due to the worldwide rice consumption, and thus, most of them is not used or used inefficiently in low value applications, for which any solution particularly like the presented with in this

Fig. 1 **a** Rice husk as obtained from the rice agricultural industry; **b** and **c** SEM images of the rice husk after the grinding process showing particles with a distribution between 1 and 20 μm

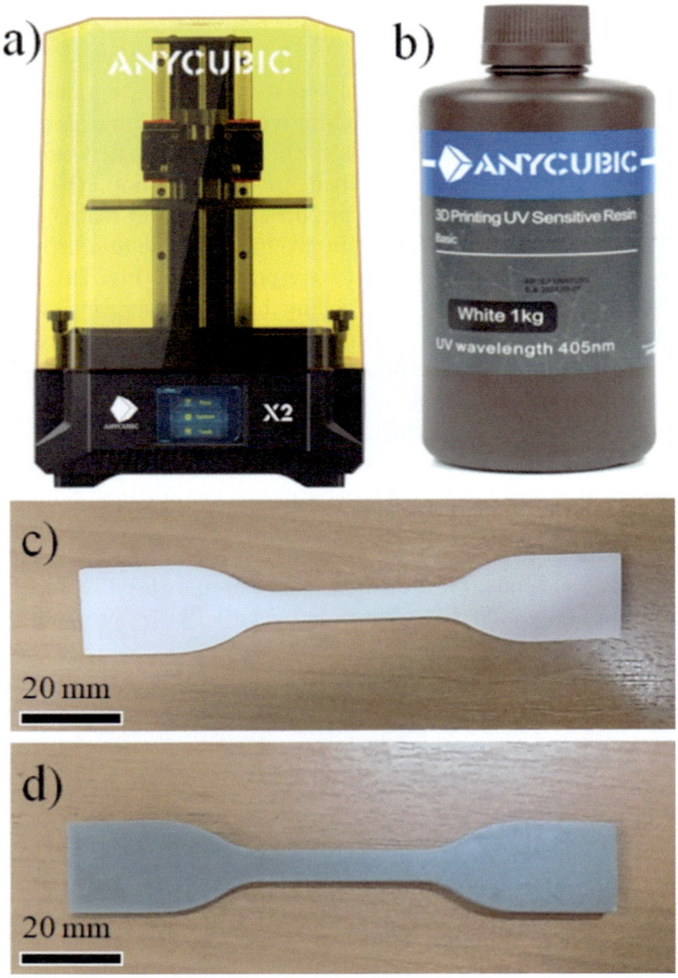

Fig. 2 **a** Printed test tube with 100% resin, **b** printed test tube with 95% resin and 5% rice husk

research that enables the material to be used in the manufacturing of parts, gives a possibility to people to create new business, and thus contributing in a sustainable way to economy.

These results could be expanded and investigated to other natural fibers, including hair waste [17], individual vegetable fibers [18], or even natural fabrics [19]. This could power economy and innovation in small communities, even in ancient communities that have plenty of natural materials only used in crafts and arts.

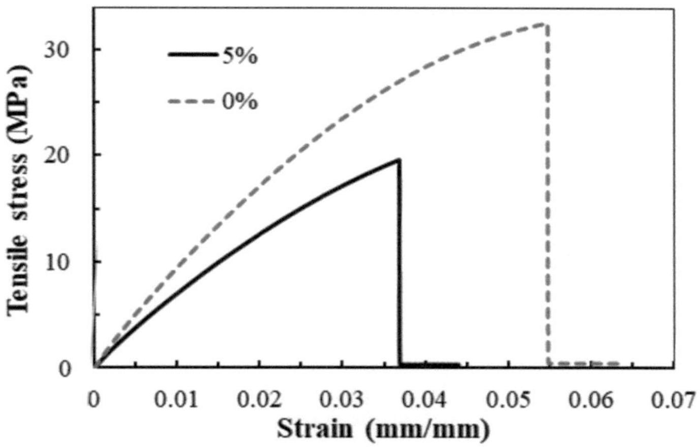

Fig. 3 Stress versus strain graph of resin specimens with rice husk fibers

Conclusions

The presented research shows that the used of rice husk is a promissory waste for vat photopolymerization, with excellent results. Additionally, the following conclusions can be summarized:

- The use of rice husk in vat photopolymerization AM is feasible to validate the process as an upcycling technology, with positive impact in both the sustainability of the rice and AM sectors.
- The addition of rice fibers to the resin makes printing difficult because it obstructs the passage of ultraviolet light, therefore requiring more research to optimize the printing layer and other parameters.
- The rice fibers improved the tensile strength and ductility as shown in Fig. 3.

References

1. Kumar S, Sangwan P, Dhankhar RMV, Bidra S (2013) Utilization of rice husk and their ash: a review. Res J Chem Env Sci 1(5):126–129
2. Zou Y, Yang T (2019) Rice husk, rice husk ash and their applications. In: Rice bran and rice bran oil. AOCS Press, pp 207–246
3. Silva LA, dos Santos IFS, de Oliveira Machado G, Tiago Filho GL, Barros RM (2021) Rice husk energy production in Brazil: an economic and energy extensive analysis. J Clean Prod 290:125188
4. Karam DS, Nagabovanalli P, Rajoo KS, Ishak CF, Abdu A, Rosli Z, Zulperi D (2022) An overview on the preparation of rice husk biochar, factors affecting its properties, and its agriculture application. J Saudi Soc Agric Sci 21(3):149–159

5. Akeke GA, Ephraim ME, Akobo IZS, Ukpata JO (2013) Structural properties of rice husk ash concrete. Int J Eng 3(3):8269
6. Colorado HA, Echeverri-Lopera GI (2020) The solid waste in Colombia analyzed via gross domestic product: towards a sustainable economy. Revista Facultad de Ingeniería Universidad de Antioquia 96:51–63
7. Agudelo G, Cifuentes S, Colorado HA (2019) Ground tire rubber and bitumen with wax and its application in a real highway. J Clean Prod 228:1048–1061
8. Hwang CL, Chandra S (1996) The use of rice husk ash in concrete. In: Waste materials used in concrete manufacturing. William Andrew Publishing, pp 184–234
9. Sharma DK, Sharma R (2018) Influence of rice husk ash and rice tiller ash along with chromate reducing agents on strength and hydration properties of ordinary Portland cement. Constr Build Mater 169:843–850
10. Suhot MA, Hassan MZ, Aziz SAA, Md Daud MY (2021) Recent progress of rice husk reinforced polymer composites: a review. Polymers 13(15):2391
11. Barreto G, Restrepo S, Vieira CM, Monteiro SN, Colorado HA (2024) Rice husk with PLA: 3D filament making and additive manufacturing of samples for potential structural applications. Polymers 16(2):245
12. Vergara LA, Perez JF, Colorado HA (2023) 3D printing of ordinary Portland cement with waste wood derived biochar obtained from gasification. Case Stud Constr Mater 18:e02117
13. Paternina Reyes MJ, Unfried Silgado J, Santa Marín JF, Colorado Lopera HA, Espitia Sanjuán LA (2023) Cashew Nutshells: a promising filler for 3D printing filaments. Polymers 15(22):4347
14. Gil LD, Monteiro SN, Colorado HA (2024) Polymer matrix nanocomposites fabricated with copper nanoparticles and photopolymer resin via vat photopolymerization additive manufacturing. Polymers 16(17):2434
15. Colorado HA, Gutierrez-Velasquez EI, Gil LD, de Camargo IL (2024) Exploring the advantages and applications of nanocomposites produced via vat photopolymerization in additive manufacturing: a review. Adv Compos Hybrid Mater 7(1):1
16. Pagac M, Hajnys J, Ma QP, Jancar L, Jansa J, Stefek P, Mesicek J (2021) A review of vat photopolymerization technology: materials, applications, challenges, and future trends of 3D printing. Polymers 13(4):598
17. Choi WS, Kim JH, Ahn CB, Lee JH, Kim YJ, Son KH, Lee JW (2021) Development of a multilayer skin substitute using human hair keratinic extract-based hybrid 3D printing. Polymers 13(16):2584
18. Deb D, Jafferson JM (2021) Natural fibers reinforced FDM 3D printing filaments. Mater Today: Proc 46:1308–1318
19. Tavares SS, de Mendonça Neuba L, Lopera HC, Monteiro SN, Meyers MA (2025) The embira bark fiber: a sustainable Amazon tape. Adv Compos Hybrid Mater 8(2):184

Author Index

H. A. C. Lopera et al. (eds.), *Natural Fibers and Biocomposites 2026*, The Minerals, Metals & Materials Series, https://doi.org/10.1007/978-3-032-13577-3

Subject Index

H. A. C. Lopera et al. (eds.), *Natural Fibers and Biocomposites 2026*, The Minerals, Metals & Materials Series, https://doi.org/10.1007/978-3-032-13577-3

MIX
Papier aus verantwortungsvollen Quellen
Paper from responsible sources
FSC® C105338

If you have any concerns about our products,
you can contact us on
ProductSafety@springernature.com

In case Publisher is established outside the EU,
the EU authorized representative is:
Springer Nature Customer Service Center GmbH
Europaplatz 3, 69115 Heidelberg, Germany

Printed by Libri Plureos GmbH
in Hamburg, Germany